Pólya Counting Theory

Combinatorics for Computer Science (Unit 4)

S. Gill Williamson

Preface

From 1970 to 1985 I ran a graduate seminar on algorithmic combinatorics in the Department of Mathematics, UCSD. Over time, I developed a series of notes or "units of study" to prepare beginning graduate students for this seminar. In 1985, these units of study were combined and published as a book, *Combinatorics for Computer Science(CCS)*, by Computer Science Press. Each of the units of study became a chapter in this book.

My general goal is to re-create the original presentation of these (largely independent) units in a form that is convenient for individual selection and study. Here, we isolate Unit 4, corresponding to Chapter 4 of *CCS*, and reconstruct the original very helpful unit specific index associated with this unit.

Theorems, figures, examples, etc., are numbered sequentially: EXERCISE 4.13 and THEOREM 4.41 refer to numbered items 13 and 41 of Unit 4 (or Chapter 4 in *CCS*).

These notes focus on the visualization of algorithms through the use of graphical and pictorial methods. This approach is both fun and powerful, preparing you to invent your own algorithms for a wide range of problems.

Basic Concepts of Linear Order is Unit 1.

Sorting and Listing is Unit 2 and 3.

S. Gill Williamson, 2012
http : \www.cse.ucsd.edu\ $\sim$ gill

Table of Contents

UNIT 4

Pólya theory and its extensions

We recall some elementary ideas from group theory. If A is a set, then a *binary operation* on A is a map β from $A \times A$ to A. We denote the element of A obtained by applying β to (x,y) by juxtaposition of x and y. Thus, $\beta(x,y) = xy$. The binary operation is *associative* if $(xy)z = x(yz)$ for all x, y, and z in A. If A is a set with an associative binary operation, then A is called a *group* if it satisfies the following two conditions:

(1) There is a fixed element e in A (called the *identity*) such that $ex = xe = x$ for all x in A.
(2) For every x in A there is an element y (called the inverse, x^{-1}, of x) such that $xy = yx = e$.

If A and B are groups and ν a mapping from A to B that satisfies the equation $\nu(xy) = \nu(x)\nu(y)$ for all x and y in A, then ν is called a *homomorphism* from A to B. If S is a set, then we denote by PER(S) the set of all permutations of S (see NOTATION 1.6). For $S = \underline{r} = \{1,2,\ldots,r\}$ or $S = \underline{r}_{\cdot} = \{0,1,\ldots,r-1\}$ we have used the notation S_r for the permutations of S. If f and g are permutations of S (i.e., elements of PER(S)), then fg is the permutation obtained by composing f and g (thus $fg(x) = f(g(x))$ for all x in S). The set of all permutations PER(S) is easily seen to be a group with this binary operation (prove it!).

4.1 DEFINITION.

Let A be a group and let S be a set. A homomorphism ν from A to PER(S), the group of permutations of S, will be called "an action of A on S." We write "A: S with ν" for "A acts on S with homomorphism ν." When there is no need to explicitly mention ν we write simply "A: S."

The study of group actions is basically the study of permutation groups. The reason for introducing the idea of a homomorphism of an abstract group is to clarify the intuitive idea of a single group having a number of different ways of acting as a permutation group, as indicated by EXAMPLE 4.2. In what follows we shall adopt our usual convention that sets are finite.

The group A associated with EXAMPLE 4.2 is the "group of symmetries" of the square. This group consists of all rotations and reflections in the plane of the square that leave the set of points representing the square invariant. Obviously, this group can be thought of as the identity, e, rotation counterclockwise by $\tau = 90°$, by $\tau^2 = 180°$, and by $\tau^3 = 270°$, and the four reflections ρ_r, ρ_s, ρ_t, ρ_u, about the axes indicated in FIGURE 4.2.

1

4.2 EXAMPLE OF A GROUP ACTION: DIHEDRAL GROUP.

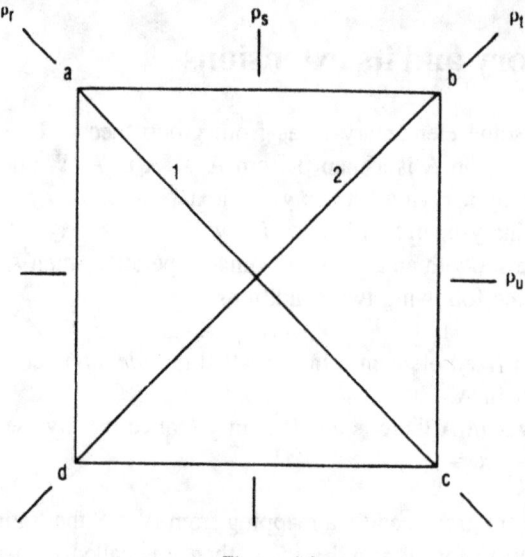

Figure 4.2

A is the group of symmetries of the square: $e, \tau, \tau^2, \tau^3, \rho_r, \rho_s, \rho_t, \rho_u$.
Action of A on vertices {a,b,c,d} defined by homomorphism ν:

$$\nu(\tau) = \begin{pmatrix} a\ b\ c\ d \\ d\ a\ b\ c \end{pmatrix}, \ \nu(\tau^2) = \begin{pmatrix} a\ b\ c\ d \\ c\ d\ a\ b \end{pmatrix},$$

$$\nu(\tau^3) = \begin{pmatrix} a\ b\ c\ d \\ b\ c\ d\ a \end{pmatrix}, \ \nu(\tau^4) = \begin{pmatrix} a\ b\ c\ d \\ a\ b\ c\ d \end{pmatrix} = \nu(e)$$

$$\nu(\rho_r) = \begin{pmatrix} a\ b\ c\ d \\ a\ d\ c\ b \end{pmatrix}, \ \nu(\rho_s) = \begin{pmatrix} a\ b\ c\ d \\ b\ a\ d\ c \end{pmatrix},$$

$$\nu(\rho_t) = \begin{pmatrix} a\ b\ c\ d \\ c\ b\ a\ d \end{pmatrix}, \ \nu(\rho_u) = \begin{pmatrix} a\ b\ c\ d \\ d\ c\ b\ a \end{pmatrix}.$$

Action of A on diagonals 1 and 2 defined by homomorphism μ:

$$\mu(\tau) = \mu(\tau^3) = \begin{pmatrix} 1\ 2 \\ 2\ 1 \end{pmatrix}, \ \mu(\tau^2) = \mu(e) = \begin{pmatrix} 1\ 2 \\ 1\ 2 \end{pmatrix},$$

$$\mu(\rho_r) = \mu(\rho_t) = \begin{pmatrix} 1\ 2 \\ 1\ 2 \end{pmatrix}, \ \mu(\rho_s) = \mu(\rho_u) = \begin{pmatrix} 1\ 2 \\ 2\ 1 \end{pmatrix}$$

Imagine now that the square of FIGURE 4.2 is rotated counterclockwise by
$\tau = 90°$. The vertex labeled a moves to where d used to be, b moves to where
a used to be, etc. We record this transformation of vertices by $\nu(\tau) =$

2

$\begin{pmatrix} a\ b\ c\ d \\ d\ a\ b\ c \end{pmatrix}$. Similarly, $\nu(\rho_r) = \begin{pmatrix} a\ b\ c\ d \\ a\ d\ c\ b \end{pmatrix}$. From the defnition of the elements

τ and ρ_r of A, it is evident that the result of applying first ρ_r and then τ, which we indicate by the product $\tau\rho_r$, is the same as the reflection ρ_u. To see this, one can think of how the group elements transform the edges of the square, the interior triangular regions, or the vertices. One can also, without thinking at all of the geometry, simply compose the two permutations $\nu(\tau)$ and $\nu(\rho_r)$ to obtain $\begin{pmatrix} a\ b\ c\ d \\ d\ c\ b\ a \end{pmatrix}$ and note that the latter permutation is $\nu(\rho_u)$. The composition, by our rules, of $\nu(\tau)$ and $\nu(\rho_r)$ is computed with the argument on the right. For example, $\nu(\tau)\nu(\rho_r)(b)$ is computed first by computing $\nu(\rho_r)(b) = d$ and then $\nu(\tau)(d) = c$.

4.3 EXERCISE.

(1) Let A be the group of symmetries of the square (FIGURE 4.2). Show that for any reflection ρ, $\tau\rho = \rho\tau^{-1}$. Show that for any fixed choice of ρ, each element of the group A can be written uniquely in the form $\tau^i\rho^j$ where $0 \leqslant i \leqslant 3$, and $0 \leqslant j \leqslant 1$.

(2) Prove that ν and μ of FIGURE 4.2 are homomorphisms. (*Hint:* The previous exercise is a help here.) In fact, ν is also a bijection. A bijective homomorphism is called an *isomorphism*.

(3) Isn't the fact that ν and μ are isomorphisms intuitively obvious? There is one detail worth considering, however. Suppose we had interpreted $\nu(\tau) = \begin{pmatrix} a\ b\ c\ d \\ d\ a\ b\ c \end{pmatrix}$ geometrically as: "a is replaced by d," "b is replaced by a," "c is replaced by b," "d is replaced by c." At first glance it again is "obvious" that ν is an isomorphism. Is it true that ν is an isomorphism with this geometric interpretation? *Hint.* Check that $\nu(\alpha\beta) = \nu(\beta)\nu(\alpha)$ so ν is an *antiisomorphism*.

4.4 NOTATION.

In most instances it is not necessary to make explicit reference to the homomorphism ν in connection with group actions. In this case, instead of writing $\nu(a)(x)$ for the image of x under $\nu(a)$, we simply write ax. If a and b are elements of A, then the fact that ν is a homomorphism is expressed by the identity (ab)x = a(bx) for all x in S (we assume A: S). **a,b are in the group now not S!**

4.5 DEFINITION.

Let A be a group and S a set with A: S. We define an equivalence relation $\sim$ on S by s $\sim$ t if there is an a $\in$ A such that as = t. The equivalence classes of this equivalence relation are called the *orbits* of A acting on S.

4.6 EXERCISE.

Prove that the relation ~ defined in DEFINITION 4.5 is actually an equivalence relation in the sense of DEFINITION 1.2.

4.7 DEFINITION.

Let A: S and let s be an element of S. The set of all elements a of A for which as = s is a subgroup A_s of A, called the *stability subgroup of* A *at* s or the *stabilizer subgroup of* A *at* s.

The orbits of a group A acting on squares with vertices labeled with two symbols are shown in FIGURE 4.8. If s is the first element shown in Orbit 4, then $A_s = \{e, \rho_u\}$ where ρ_u is as shown in FIGURE 4.2. Note that $|A| = 8$ and $|A_s| = 2$ and there are $|A|/|A_s| = 4$ elements in Orbit 4. This is a general result, stated in LEMMA 4.9.

4.8 ORBITS OF A DIHEDRAL GROUP ACTION.

Let S = the set of all functions $\{1,2,3,4\} \rightarrow \{g,r\}$, or equivalently, the set of all 2-colorings of the four vertices of a square. If A is the group of rotations and reflections of a square, then A: S has six different orbits.

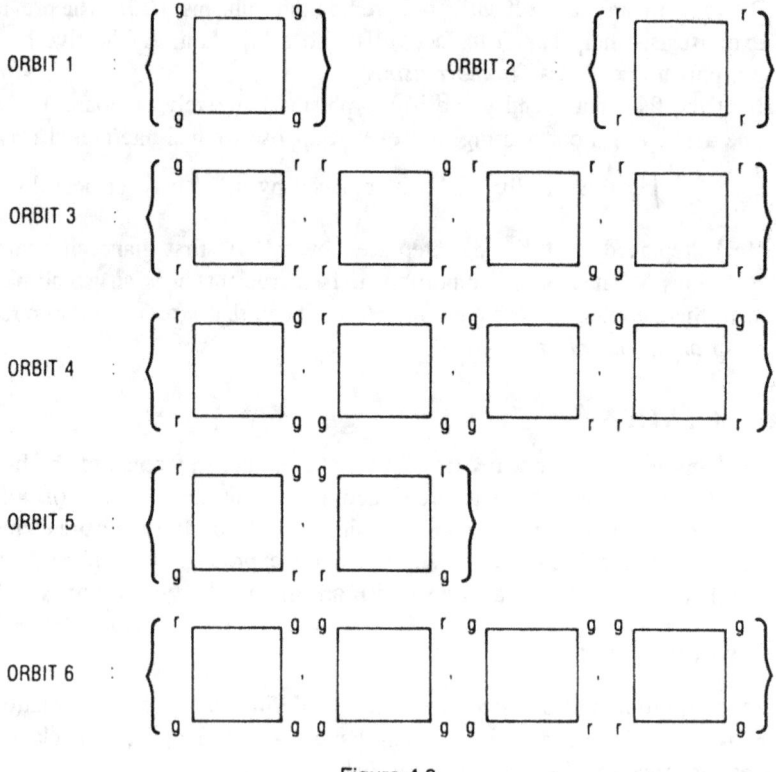

Figure 4.8

4

The reader should recall from elementary group theory that if A is a group and B is a subgroup of A, then A/B is the set $\{aB: a \in A\}$ of cosets of B in A.

4.9 LEMMA.

Let A: S and for each $s \in S$ let $O_s = \{as: a \in A\}$ be the orbit of s. Let $A/A_s = \{aA_s: a \in A\}$ be the set of cosets of A_s in A. Then the map f: $O_s \rightarrow A/A_s$ defined by $f(as) = aA_s$ is a bijection. Hence, $|O_s| = |A|/|A_s|$.

Proof. Observe that for a and b in A and s in S, we have as = bs if and only if $b^{-1}as = s$ if and only if $b^{-1}a \in A_s$ if and only if $aA_s = bA_s$. Reading this chain of implications from left to right shows that f is a function, and reading from right to left shows that f is an injection. It is obvious that f is a surjection, hence f is a bijection.

It is immediate from LEMMA 4.9 that if $s \sim t$ (i.e., s and t lie in the same orbit) then $|A_s| = |A_t|$. Again, from elementary group theory, recall that if A is a group with subgroups B and C, then B and C are said to be *conjugate subgroups* of A if there is an element $a \in A$ such that $aBa^{-1} = C$. LEMMA 4.10 asserts that elements in the same orbit of A: S have conjugate stabilizer subgroups.

4.10 LEMMA.

Let A: S and suppose that for s and t in S there is an a in A such that as = t. Then $aA_s a^{-1} = A_t$.

Proof. Note that bt = t if and only if b(as) = as if and only if $a^{-1}bas = s$ if and only if $a^{-1}ba$ is in A_s. Thus, $A_s = a^{-1}A_t a$.

If A is a group, then we can use the idea of conjugation to define an action A: A. For a in A and x in A define $\nu(a)x = axa^{-1}$. Then it is easily seen that ν is a homomorphism from A to PER(A). We say that "A acts on itself by conjugation." The orbits of this action are called the *conjugacy classes* of A. If x is in A, then the stabilizer subgroup of x in A under conjugation, $\{a: axa^{-1} = x\}$, is denoted by C_x and is called the *centralizer* of x in A. If K is the conjugacy class containing x then, by LEMMA 4.9, $|K| = |A|/|C_x|$.

We now discuss a class of results that are related to "Burnside's Lemma" in group theory. Let A: S be a group action. We order the group elements A and also order the elements of the set S. A matrix M whose entries M(a,s) are indexed by the pairs $A \times S$ will be called an *action matrix* for A: S. An action matrix for the case where A is the group of rotations and reflections of the triangle and S is the set of triangles with vertices labeled either r or g is shown in FIGURE 4.11. In FIGURE 4.11, M(a,s) = 1 if as = s and M(a,s) = 0 if as $\neq$ s. In general, an action matrix M(a,s) that satisfies M(a,s) = 0 if as $\neq$ s will be called *stable*. Thus, the action matrix of FIGURE 4.11 is stable. If it is the case that whenever a and b are conjugate elements of A then the row sums $\sum_{s \in S} M(a,s)$ and $\sum_{s \in S} M(b,s)$ are equal, then M will be called *class consistent*. Similarly, if s and t in the

5

same orbit of A: S implies that the column sums $\sum_{a\in A} M(a,s)$ and $\sum_{a\in A} M(a,t)$ are equal, then M will be called *orbit consistent*. Assume addition and subtraction of entries of M are defined (see the examples that follow).

Note that the action matrix of FIGURE 4.11 is both orbit and class consistent. In fact, $\sum_{a\in A} M(a,s) = |A_s|$, the size of the stabilizer subgroup of A at s. Thus, orbit consistency follows from LEMMA 4.10. The row sum $\sum_{s\in S} M(a,s) = |S_a|$ where $S_a = \{s: s \in S, as = s\}$. The reader should verify that if a,b, and c are elements of A with $c^{-1}bc = a$ (i.e., a and b are conjugate elements of A), then $cS_a = S_b$. This fact implies immediately that the matrix M of FIGURE 4.11 is class consistent.

We have, in fact, two actions of the group A. The group A acts on S and also acts on itself by conjugation. Thus, A acts on A × S by the rule, $a(x,s) = (axa^{-1},as)$ where $x \in A$ and $s \in S$. We call this action the *product action of* A *on* A × S.

4.11 ACTION MATRIX FOR ROTATIONS AND REFLECTIONS OF A TRIANGLE.

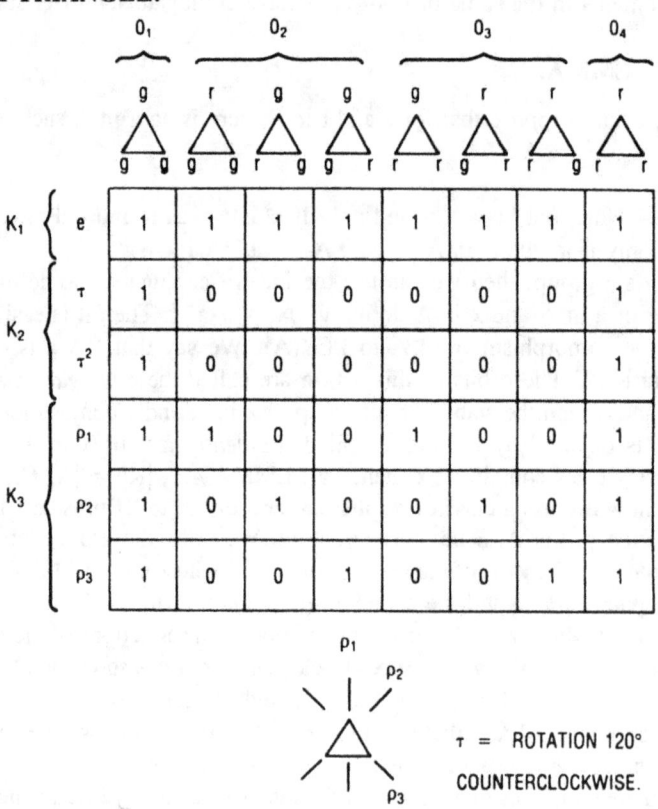

Figure 4.11

4.12 DEFINITION.

Let A: S. A stable action matrix M will be called a *Burnside* matrix if $M(x,s) = M(axa^{-1},as)$ for all a and x in A and s in S. In other words, an action matrix M that is stable ($M(x,s) = 0$ if $xs \neq s$) and is constant on orbits of the product action of A on $A \times S$ is a Burnside matrix.

4.13 EXERCISE.

(1) Prove that the "product action" of A on $A \times S$ as defined above is a group action in the sense of DEFINITION 4.1.
(2) Prove that any action matrix M (stable or not) that satisfies the equation $M(x,s) = M(axa^{-1},as)$ for all $a,x \in A$ and $s \in S$ is orbit and class consistent as defined in the discussion following LEMMA 4.10.

4.14 DEFINITION.

Let A: S be a group action. For $a \in A$, $S_a = \{s: s \in S, as = s\}$ is the subset of S stabilized by a. For $s \in S$, $A_s = \{a: a \in A, as = s\}$ is the stabilizer subgroup of A at s. $C_x = \{a: a \in A, axa^{-1} = x\}$ is the centralizer of x in A. Let $\Delta(S)$ be a system of representatives for the orbits of A: S. Let $\Delta(A)$ be a system of representatives for the conjugacy classes of A (the orbits of A: A by conjugation).

4.15 LEMMA (generalized Burnside's lemma).

Let A: S be a group action with action matrix M. If M is a Burnside matrix, then the following are equal:

(1) $\displaystyle\sum_{s \in S} \sum_{a \in A_s} M(a,s)$

(2) $\displaystyle\sum_{a \in A} \sum_{s \in S_a} M(a,s)$

(3) $\displaystyle\sum_{s \in \Delta(S)} \frac{|A|}{|A_s|} \sum_{a \in A_s} M(a,s)$

(4) $\displaystyle\sum_{a \in \Delta(A)} \frac{|A|}{|C_a|} \sum_{s \in S_a} M(a,s)$.

Proof. Using the fact that $M(a,s) = 0$ if $as \neq s$ (i.e., M is stable), it is obvious that (1) equals (2) because both equal the sum of all of the entries in M. By EXERCISE 4.13(2), (1) equals (3) since $|A|/|A_s|$ is the number of elements in the orbit of s. Similarly, (2) equals (4) since $|A|/|C_a|$ is the number of elements in the conjugacy class of a.

LEMMA 4.15 is an important and very basic tool in many different types of orbit enumeration problems. The action matrix M is more a conceptual tool than a practical one. Conceptually, M is used to give one insights into how to apply LEMMA 4.15. The entries of M can be numbers, variables, polynomials, formal power series, and even the entries of A and S themselves (the symbols of A and S can be regarded as variables in polynomials and formal power series). We shall now give a number of examples of the application of LEMMA 4.15. In the case where $M(a,s) = W(s)$ does not depend on $a \in A$, LEMMA 4.15 takes on a particularly simple form which we state in COROLLARY 4.16.

4.16 COROLLARY.

If $M(a,s) = W(s)$ does not depend on the group element a, we have:

(1) $\displaystyle \sum_{s \in \Delta(S)} W(s) = \frac{1}{|A|} \sum_{a \in A} \left(\sum_{s \in S_a} W(s) \right)$

  (weighted Burnside's lemma)

(2) $\displaystyle |\Delta(S)| = \frac{1}{|A|} \sum_{a \in A} |S_a|$

  (classical Burnside's lemma)

Proof. Identity (1) follows by setting 4.15 (3) equal to 4.15 (2) and dividing by $|A|$. Identity (2) follows from (1) by setting $W(s) = 1$ for all s.

4.17 ACTION MATRIX FOR EXAMPLE 4.18 (see FIGURE 4.20 for symmetries of triangle)

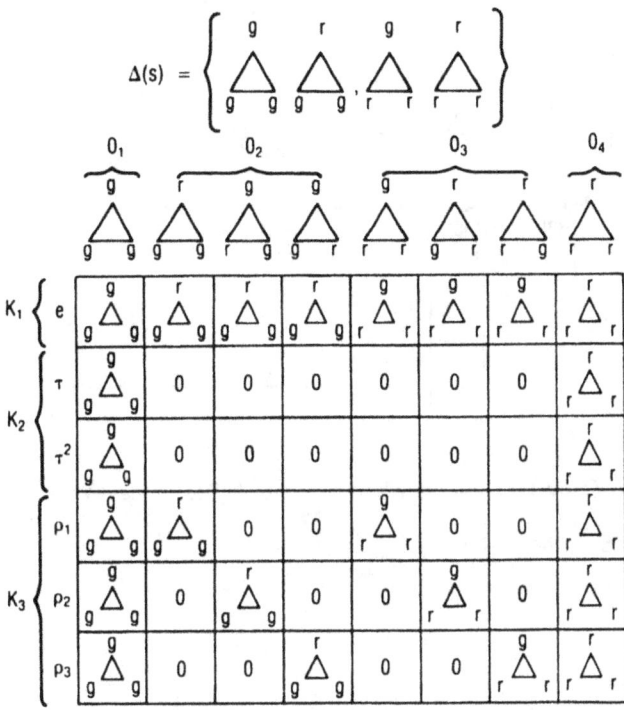

Figure 4.17

As our first example, we consider the "weighted Burnside's lemma" of COROLLARY 4.16(1). The action matrix for this example is shown in FIGURE 4.17. The group is the group of symmetries of a triangle shown in Figure 4.20. We take $W(s') = s$ where s is the unique element of $\Delta(S)$ equivalent to s'. The set of representatives $\Delta(s)$ is shown in FIGURE 4.17. It may seem odd to the reader to take sums of labeled triangles as is done in EXAMPLE 4.18. The easiest way to adjust to this idea is to think of the labeled triangles (i.e., the elements of S in general) as symbols representing variables in a polynomial (a linear polynomial in this case). The reader would not be shocked to see a polynomial in variables x, y, and z or other letters of the alphabet. If our ancestors had chosen to represent the letters of the alphabet with labeled triangles instead of the present symbols then polynomials such as those in EXAMPLE 4.18 would have seemed perfectly all right! The reader should look at EXAMPLE 4.18 in detail but should also interpret the result directly in terms of the action matrix of FIGURE 4.17. Basically, the result says that each element of $\Delta(S)$ occurs exactly $|A| = 6$ times in the action matrix and expresses this fact in terms of row sums.

9

4.18 EXAMPLE OF BURNSIDE'S LEMMA AS MANIPULATION OF SET ELEMENTS.

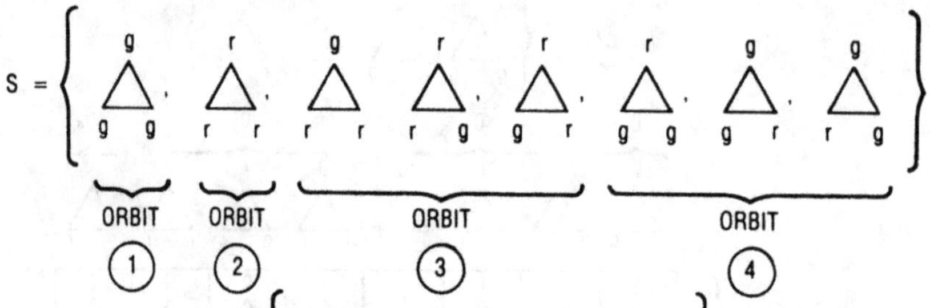

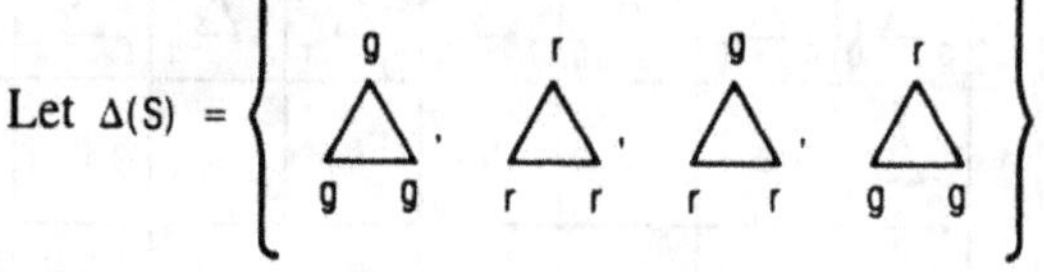

$$S = \left\{ \triangle, \triangle, \triangle, \triangle, \triangle, \triangle, \triangle, \triangle \right\}$$

ORBIT ① ORBIT ② ORBIT ③ ORBIT ④

$$\text{Let } \Delta(S) = \left\{ \triangle, \triangle, \triangle, \triangle \right\}$$

Define O by $O(x) = s$, whenever $s \in \Delta(S)$, $x \sim s$. Verify that $|A| \sum_{s \in \Delta} s = \sum_{a \in A} O(S_a)$ where $O(S_a) = \sum_{x \in S_a} O(x)$.

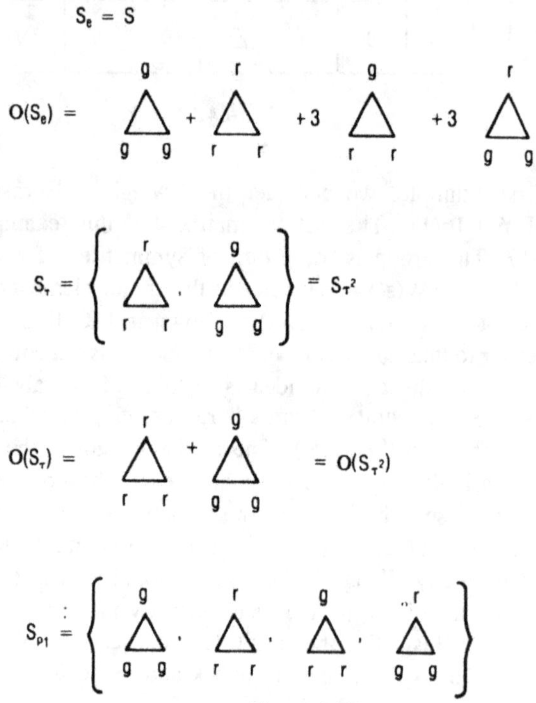

$$S_e = S$$

$$O(S_e) = \triangle + \triangle + 3\,\triangle + 3\,\triangle$$

$$S_\tau = \left\{ \triangle, \triangle \right\} = S_{\tau^2}$$

$$O(S_\tau) = \triangle + \triangle = O(S_{\tau^2})$$

$$S_{p_1} = \left\{ \triangle, \triangle, \triangle, \triangle \right\}$$

Figure 4.18

(cont.)

10

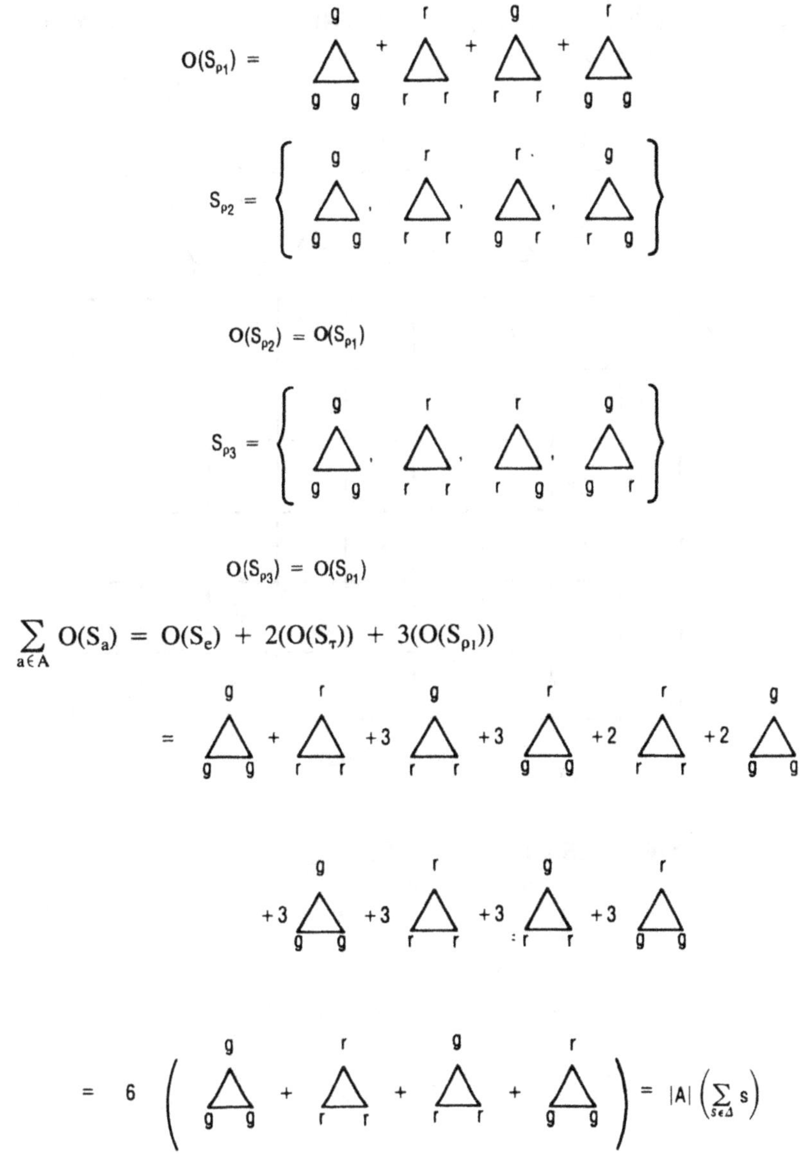

$$O(S_{p_1}) = \underset{g\quad g}{\overset{g}{\triangle}} + \underset{r\quad r}{\overset{r}{\triangle}} + \underset{r\quad r}{\overset{g}{\triangle}} + \underset{g\quad g}{\overset{r}{\triangle}}$$

$$S_{p_2} = \left\{ \underset{g\quad g}{\overset{g}{\triangle}} ,\ \underset{r\quad r}{\overset{r}{\triangle}} ,\ \underset{g\quad r}{\overset{r}{\triangle}} ,\ \underset{r\quad g}{\overset{g}{\triangle}} \right\}$$

$$O(S_{p_2}) = O(S_{p_1})$$

$$S_{p_3} = \left\{ \underset{g\quad g}{\overset{g}{\triangle}} ,\ \underset{r\quad r}{\overset{r}{\triangle}} ,\ \underset{r\quad g}{\overset{r}{\triangle}} ,\ \underset{g\quad r}{\overset{g}{\triangle}} \right\}$$

$$O(S_{p_3}) = O(S_{p_1})$$

$$\sum_{a \in A} O(S_a) = O(S_e) + 2(O(S_\tau)) + 3(O(S_{p_1}))$$

$$= \underset{g\quad g}{\overset{g}{\triangle}} + \underset{r\quad r}{\overset{r}{\triangle}} + 3\underset{r\quad r}{\overset{g}{\triangle}} + 3\underset{g\quad g}{\overset{r}{\triangle}} + 2\underset{r\quad r}{\overset{r}{\triangle}} + 2\underset{g\quad g}{\overset{g}{\triangle}}$$

$$+ 3\underset{g\quad g}{\overset{g}{\triangle}} + 3\underset{r\quad r}{\overset{r}{\triangle}} + 3\underset{r\quad r}{\overset{g}{\triangle}} + 3\underset{g\quad g}{\overset{r}{\triangle}}$$

$$= 6 \left(\underset{g\quad g}{\overset{g}{\triangle}} + \underset{r\quad r}{\overset{r}{\triangle}} + \underset{r\quad r}{\overset{g}{\triangle}} + \underset{g\quad g}{\overset{r}{\triangle}} \right) = |A| \left(\sum_{s \in \Delta} s \right)$$

Figure 4.18
(continued)

11

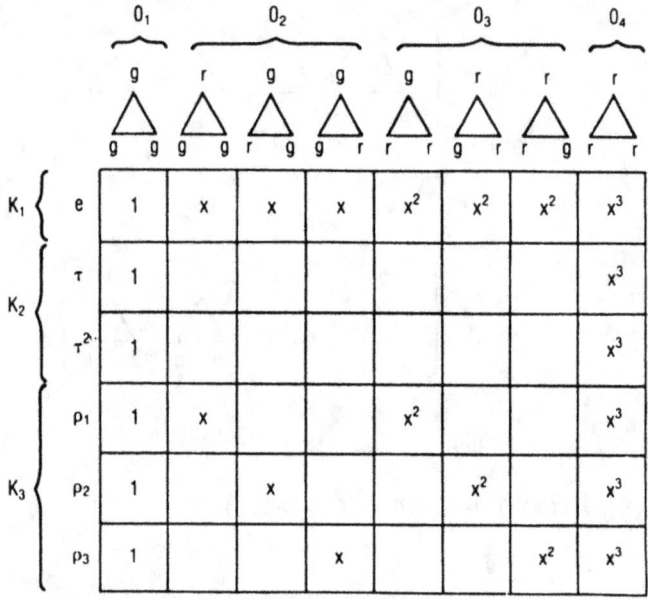

		O_1	O_2			O_3			O_4
		g	r	g	g	g	r	r	r
		g	g	g	r	r	r	g	r
K_1 {	e	1	x	x	x	x^2	x^2	x^2	x^3
K_2 {	τ	1							x^3
	τ^2	1							x^3
K_3 {	ρ_1	1	x			x^2			x^3
	ρ_2	1		x			x^2		x^3
	ρ_3	1			x			x^2	x^3

Figure 4.19

4.20 SYMMETRIES OF A TRIANGLE.

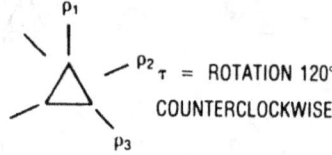

τ = ROTATION 120° COUNTERCLOCKWISE.

Figure 4.20

FIGURE 4.17 represents, in a certain sense, the most general possible type of stable action matrix where M(a,s) depends only on s. As EXAMPLE 4.18 shows, the weighted Burnside's lemma for this general case is a statement about what happens when the elements of the set S are put in stacks according to stabilizer elements. Other weaker statements can be obtained by replacing the objects themselves by various algebraic quantities. Thinking of the objects as variables, this amounts to making a change of variable in the resulting polynomials. In FIGURE 4.19, each object s is replaced by the expression $x^{r(s)}$ where

12

r(s) is the number of times the symbol r occurs in the object s. In this case, the identity 4.16(1) becomes the expression $1 + x + x^2 + x^3 = (1/6)$ $(6 + 2x + 2x + 2x + 2x^2 + 2x^2 + 2x^2 + 6x^3)$. In more complex examples, the expression for the right-hand side of identity 4.16(1) is easier to compute than the left-hand side of 4.16(1). In this case, both sides are trivial. The expression $1 + x + x^2 + x^3$ tells us that there is one element of $\Delta(S)$ with no r's, one with one r, one with two r's, and one with three r's. For more complex examples this type of information is not obvious.

4.21 EXAMPLE OF THE CLASSICAL BURNSIDE'S LEMMA.

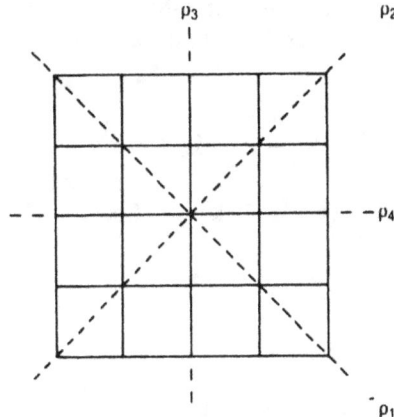

Figure 4.21

Given a square array of 16 squares, how many different ways are there to put 8 x's into these 16 squares, up to rotations and reflections? Let τ be counterclockwise rotation by 90° and let ρ_1, ρ_2, ρ_3, and ρ_4 be the reflections about the axes indicated in the above figure. Let $A = \{e, \tau, \tau^2, \tau^3, \rho_1, \rho_2, \rho_3, \rho_4\}$. Using 4.16(2) we see that

$$|\Delta(S)| = 1/|A| \sum_{a \in A} |S_a|$$

$$= 1/8 \left(|S_e| + |S_\tau| + |S_{\tau^2}| + |S_{\tau^3}| + |S_{\rho 1}| \right.$$

$$\left. + |S_{\rho 2}| + |S_{\rho 3}| + |S_{\rho 4}| \right)$$

$$= 1/8 \left(\binom{16}{8} + \binom{4}{2} + \binom{8}{4} + \binom{4}{2} \right.$$

$$\left. + 2 \left(\binom{6}{4} + \binom{4}{2}\binom{6}{3} + \binom{6}{2} \right) + 2 \binom{8}{4} \right) = 1,674.$$

13

For our next example we consider an action matrix M in which M(a,s) depends on both a and s. Let A be any group. Suppose $\lambda: A \to \mathbb{C}$ is a complex valued group homomorphism; i.e., $\lambda(aa') = \lambda(a)\lambda(a')$ for all $a, a' \in A$. If A' is a subgroup of A, then what possible values can $\sum_{a' \in A'} \lambda(a')$ have? Choose any $x \in A'$. Then $\lambda(x) \sum_{a' \in A'} \lambda(a') = \sum_{a' \in A'} \lambda(xa') = \sum_{a \in A'} \lambda(a)$. If we let $\beta = \sum_{a' \in A'} \lambda(a')$, this says that if $x \in A'$, $\lambda(x)\beta = \beta$. This implies that either $\beta = 0$, or $\lambda(x) = 1$ for all $x \in A'$ (i.e., $\beta = |A'|$). Hence $\lambda(A') = \sum_{a' \in A'} \lambda(a') = |A'|$ or 0. The former case occurs only if $A' \subseteq \ker\lambda = \{a: a \in A, \lambda(a) = 1\}$. Assume A: S. We define $M(x,s) = 0$ if $xs \neq s$ and $M(x,s) = \lambda(x)O(s)$ if $xs = s$ (O(s) as in EXAMPLE 4.18). It is easily seen that M so defined is a Burnside matrix as $M(axa^{-1},as) = \lambda(axa^{-1})O(as) = \lambda(a)\lambda(x)\lambda(a^{-1})O(s) = \lambda(x)O(s) = M(x,s)$. Thus, 4.15(3) becomes

$$\sum_{s \in \Delta(S)} \frac{|A|}{|A_s|}\left(\sum_{a \in A_s} \lambda(a)\right) s = |A| \sum_{s \in \hat{\Delta}(S)} s$$

where $\hat{\Delta}(S) = \{s: s \in \Delta(S), A_s \subseteq \ker\lambda\}$. Equating 4.15(2) and 4.15(3) gives IDENTITY 4.22.

4.22 IDENTITY.

$$\sum_{s \in \hat{\Delta}(S)} s = \frac{1}{|A|} \sum_{a \in A} \lambda(a)O(S_a).$$

IDENTITY 4.22 forms the basis for EXAMPLE 4.23 and EXAMPLE 4.24.

4.23 BURNSIDE'S LEMMA WITH A GROUP CHARACTER.

Let S = all functions from $\{1,2,3,4\} \to \{g,r\}$. Let $A = \{e,\tau,\tau^2,\tau^3\}$ where τ is a 90° rotation counterclockwise. S has six orbits under A. Let

$$\Delta = \left\{ \begin{smallmatrix} g & & g \\ & \square & \\ g & & g \end{smallmatrix}, \begin{smallmatrix} g & & g \\ & \square & \\ g & & r \end{smallmatrix}, \begin{smallmatrix} g & & g \\ & \square & \\ r & & r \end{smallmatrix}, \begin{smallmatrix} g & & r \\ & \square & \\ r & & r \end{smallmatrix}, \begin{smallmatrix} r & & r \\ & \square & \\ r & & r \end{smallmatrix}, \begin{smallmatrix} g & & r \\ & \square & \\ r & & g \end{smallmatrix} \right\}.$$

Let $\lambda = \begin{pmatrix} e & \tau & \tau^2 & \tau^3 \\ 1 & i & -1 & -i \end{pmatrix}$. We verify that $1/|A| \sum_{a \in A} \lambda(a)O(S_a) = \sum_{s \in \hat{\Delta}(S)} O(s)$, where

$O(x) = s$ if $s \in \Delta$ and $x \sim s$. $\hat{\Delta}(S) = \{s: s \in \Delta, A_s \subseteq \ker\lambda\} =$

$$\left\{ \begin{smallmatrix} g & & g \\ & \square & \\ g & & r \end{smallmatrix}, \begin{smallmatrix} g & & r \\ & \square & \\ r & & r \end{smallmatrix}, \begin{smallmatrix} g & & g \\ & \square & \\ r & & r \end{smallmatrix} \right\}.$$

Compute $\lambda(a)O(S_a)$ for each $a \in A$:

$$\lambda(e)O(S_e) = 1\,O(S)$$

$$= \overset{g\ g}{\underset{g\ g}{\square}} + 4\,\overset{g\ g}{\underset{g\ r}{\square}} + 4\,\overset{g\ g}{\underset{r\ r}{\square}} + 4\,\overset{g\ r}{\underset{r\ r}{\square}}$$

$$+ 2\,\overset{g\ r}{\underset{r\ g}{\square}} + \overset{r\ r}{\underset{r\ r}{\square}}$$

$$\lambda(\tau)O(S_\tau) = i\,\overset{g\ g}{\underset{g\ g}{\square}} + i\,\overset{r\ r}{\underset{r\ r}{\square}}$$

$$\lambda(\tau^2)O(S_{\tau^2}) = -1\,\overset{g\ g}{\underset{g\ g}{\square}} - 2\,\overset{g\ r}{\underset{r\ g}{\square}} - 1\,\overset{r\ r}{\underset{r\ r}{\square}}$$

$$\lambda(\tau^3)O(S_{\tau^3}) = -i\,\overset{g\ g}{\underset{g\ g}{\square}} - i\,\overset{r\ r}{\underset{r\ r}{\square}}$$

$$\frac{1}{|A|}\sum_{a\in A}\lambda(a)O(S_a) = \frac{1}{4}\left(4\,\overset{g\ g}{\underset{g\ r}{\square}} + 4\,\overset{g\ g}{\underset{r\ r}{\square}} + 4\,\overset{g\ r}{\underset{r\ r}{\square}}\right)$$

$$= \overset{g\ g}{\underset{g\ r}{\square}} + \overset{g\ g}{\underset{r\ r}{\square}} + \overset{g\ r}{\underset{r\ r}{\square}} = \sum_{s\in\hat{\Delta}(S)} O(s).$$

4.24 SECOND EXAMPLE OF BURNSIDE'S LEMMA AND A GROUP CHARACTER.

Let S, A, and Δ be as in EXAMPLE 4.23, but let $\lambda = \begin{pmatrix} e & \tau & \tau^2 & \tau^3 \\ +1 & -1 & +1 & -1 \end{pmatrix}$.

We show that $\frac{1}{|A|}\sum_{a\in A}\lambda(a)O(S_a) = \sum_{s\in\hat{\Delta}(S)} O(s)$, where $\hat{\Delta}(S) = \{s: s \in \Delta, A_s \subseteq \ker\lambda\} = \{s: s \in \Delta, A_s \subseteq \{e,\tau^2\}\}$

$$= \left\{ \overset{g\ g}{\underset{g\ r}{\square}}, \overset{g\ g}{\underset{r\ r}{\square}}, \overset{g\ r}{\underset{r\ r}{\square}}, \overset{g\ r}{\underset{r\ g}{\square}} \right\}$$

$$\lambda(e)O(S_e) = \overset{g\ g}{\underset{g\ g}{\square}} + 4\,\overset{g\ g}{\underset{g\ r}{\square}} + 4\,\overset{g\ g}{\underset{r\ r}{\square}} + 4\,\overset{g\ r}{\underset{r\ r}{\square}}$$

$$+ 2\,\overset{g\ r}{\underset{r\ g}{\square}} + \overset{r\ r}{\underset{r\ r}{\square}}$$

$$\lambda(\tau)O(S_\tau) = - \overset{g\ g}{\underset{g\ g}{\square}} - \overset{r\ r}{\underset{r\ r}{\square}} = \lambda(\tau^3)O(S_{\tau^3})$$

$$\lambda(\tau^2)O(S_{\tau^2}) = \overset{g\ g}{\underset{g\ g}{\square}} + 2\,\overset{g\ r}{\underset{r\ g}{\square}} + \overset{r\ r}{\underset{r\ r}{\square}}. \text{ So}$$

15

4.25 IDENTITY.

$$\frac{1}{|A|}\sum_{a\in A}\lambda(a)O(S_a) = \frac{1}{4}\left(4\,{}^{g}\square^{g}_{g}{}_{r} + 4\,{}^{g}\square^{g}_{r}{}_{r} + 4\,{}^{g}\square^{r}_{r}{}_{r} + 4\,{}^{g}\square^{r}_{r}{}_{g}\right)$$

$$= {}^{g}\square^{g}_{g}{}_{r} + {}^{g}\square^{g}_{r}{}_{r} + {}^{g}\square^{r}_{r}{}_{r} + {}^{g}\square^{r}_{r}{}_{g}$$

$$= \sum_{s\in\hat{\Delta}(S)} O(s).$$

4.26 EXERCISE.

(1) Use the ideas associated with LEMMA 4.15 and COROLLARY 4.16 to discuss EXERCISE 1.38(4) and (5).

(2) Use the ideas associated with LEMMA 4.15 and COROLLARY 4.16 to discuss the domino covering problem of FIGURE 1.37, EXERCISE 1.38(2) and (3). Try to analyze some larger boards than the 4 × 4 board.

(3) Use the ideas associated with LEMMA 4.15 and COROLLARY 4.16 to analyze the problem of covering a 2 × n board with dominoes.

We conclude our discussion of Burnside's lemma and its variations by proving White's lemma on orbit representatives with conjugate stability subgroups.

4.27 NOTATION.

We define $\chi(statement) = 1$ if $statement$ is true and $\chi(statement) = 0$ if $statement$ is false.

Consider $\sum_{s\in S} W(s)\chi(H \subseteq A_s)$ where $W(s)$ is as in COROLLARY 4.16. This sum represents all values of W for elements of s whose stability subgroup contains a fixed group H as a subgroup. For many examples, this sum is not too difficult to calculate and, so it seems, is also not very interesting! A much more interesting and difficult sum to compute is $\sum_{s\in\Delta(S)} W(s)\chi(A_s \sim K)$ where K is a fixed subgroup of A. This sum represents all elements s in $\Delta(S)$ (DEFINITION 4.14) whose stability subgroups A_s are conjugate to a fixed subgroup K (the notation $A_s \sim K$ means "A_s is conjugate to K"). There is a remarkable relationship between these two sums that represents an important combinatorial technique (one that we shall see more of in Chapter 5). Using the fact that W is constant on orbits

of A: S, we see that $\sum_{s\in S} W(s)\chi(H \subseteq A_s) = \sum_{s\in\Delta(S)} W(s)\left(\sum_{s'\sim s} \chi(H \subseteq A_{s'})\right)$. Note

that $\sum_{s'\sim s} f(s') = \frac{1}{|A_s|}\sum_{a\in A} f(as)$ for any function f so that above sum becomes

$$\sum_{s \in \Delta(S)} \frac{W(s)}{|A_s|} \left(\sum_{a \in A} \chi(H \subseteq A_{as}) \right) \quad \text{which, from LEMMA 4.10, equals}$$

$\sum_{s \in \Delta(S)} \frac{W(s)}{|A_s|} \left(\sum_{a \in A} \chi(a^{-1}Ha \subseteq A_s) \right)$. If we let $\mathcal{A}$ be the set of all subgroups of A, then A acts on $\mathcal{A}$ by conjugation ($K \in \mathcal{A}$ is sent to aKa^{-1}). Let $\mathcal{H}$ be a list of orbit representatives for this action (a complete list of nonconjugate subgroups). Using the fact that $\sum_{a \in A} \chi(a^{-1}Ha \subseteq A_s) = \sum_{a \in A} \chi(aHa^{-1} \subseteq A_s)$. We can now

rewrite the above expression as $\displaystyle\sum_{K \in \mathcal{H}} \sum_{s \in \Delta} \frac{W(s)}{|A_s|} \chi(A_s \sim K) \sum_{a \in A} \chi(aHa^{-1} \subseteq A_s)$

$= \displaystyle\sum_{K \in \mathcal{H}} \left(\frac{1}{|K|} \sum_{a \in A} \chi(aHa^{-1} \subseteq K) \right) \sum_{s \in \Delta(S)} W(s)\chi(A_s \sim K)$. To get this expression

we have simply reorganized the sum over $s \in \Delta \equiv \Delta(s)$ according to the element $K \in \mathcal{H}$ to which A_s is conjugate. We use the fact that $\sum_{a \in A} \chi(aHa^{-1} \subseteq A_s)$

$= \displaystyle\sum_{a \in A} \chi(aHa^{-1} \subseteq K)$ if A_s is conjugate to K. These observations lead one to

DEFINITION 4.28 and THEOREM 4.29.

4.28 DEFINITION.

Let H and K be subgroups of A. We define $M_K(H)$, called the *mark of* K *at* H, to be the sum

$$\frac{1}{|K|} \sum_{a \in A} \chi(aHa^{-1} \subseteq K).$$

We have thus proved LEMMA 4.29.

4.29 LEMMA (White's lemma).

Let W be constant on orbits of A: S. Then

$$\sum_{K \in \mathcal{H}} M_K(H) \sum_{s \in \Delta(S)} W(s)\chi(A_s \sim K) = \sum_{s \in S} W(s)\chi(H \subseteq A_s).$$

It is conceptually useful to interpret White's lemma in matrix terms. Order the complete list of nonconjugate subgroups $K_1, K_2, \ldots, K_p$ such that $|K_1| \geq |K_2| \geq \ldots \geq |K_p|$. Define a $p \times p$ matrix $\mathcal{M}$ whose entry in the position i,j is $M_{K_j}(K_i)$. Obviously, $\mathcal{M}$ is lower triangular with nonzero diagonal entries.

4.30 DEFINITION.

The nonsingular lower triangular matrix $\mathcal{M} = (M_{K_j}(K_i))$, where $|K_1| \geq |K_2| \geq \ldots \geq |K_p|$ is a complete list of nonconjugate subgroups of A, is called the *matrix of marks* of A.

Let X denote the p $\times$ 1 column vector with j^{th} entry $\sum\limits_{s \in \Delta(S)} W(s)X(A_s \sim K_j)$
and let Y denote the p $\times$ 1 column vector with i^{th} entry $\sum\limits_{s \in S} W(s)X(K_i \subseteq A_s)$.
Then IDENTITY 4.31 gives the matrix version of White's lemma.

4.31 IDENTITY (Matrix version of White's lemma).

$$\mathcal{M}X = Y \text{ and } X = \mathcal{M}^{-1}Y.$$

Given $\mathcal{M}$, the matrix $\mathcal{M}^{-1}$ is easily computed since $\mathcal{M}$ is lower triangular. The matrix $\mathcal{M}$ itself is not generally easy to compute. It is possible to compute $\mathcal{M}$ for certain important crystallographic groups. For some interesting applications of White's lemma, see the articles by McLarnan and McLarnan and Moore cited in the references at the end .

4.32 EXERCISE.

(1) Let A be a group with subgroups K and H. Show that the mark of K at H, $M_K(H)$, is given by the formula $\dfrac{|A|\ [H \subseteq K]}{|K|\ [H]}$ where [H $\subseteq$ K] denotes the number of subgroups of A conjugate to H and contained in K and [H] = [H $\subseteq$ A].

(2) Calculate the matrix of marks for the symmetry groups of the triangle and square.

(3) TABLE 4.33 gives the list of nonconjugate subgroups for the rotations and reflections of the hexagon. The matrix of marks for this group is also given in TABLE 4.33. Compute the vector X of IDENTITY 4.31 for the case where S is the set of all hexagons labeled with two symbols g and r. Do the same when S is the set of labeled hexagons of EXERCISE 1.38(5).

4.33 MATRIX OF MARKS FOR THE SYMMETRY GROUP OF THE HEXAGON.

K_1 = G;

K_2 = cyclic group of order 6;

K_3 = dihedral group of order 6 generated by 120° rotation and reflection through line through opposite vertices;

K_4 = dihedral group of order 6 generated by 120° rotation and reflection through line through opposite midpoints;

K_5 = dihedral group of order 4 generated by 180° rotation and a reflection;

K_6 = cyclic group of order 3;

K_7 = cyclic group of order 2 generated by 180° rotation;

K_8 = cyclic group of order 2 generated by reflection through line through opposite vertices;

18

K_9 = cyclic group of order 2 generated by reflection through line through opposite midpoints;

K_{10} = identity.

$$\mathcal{M} = \begin{bmatrix} 1 & & & & & & & & \\ 1 & 2 & & & & & & & \\ 1 & 0 & 2 & & & & 0 & & \\ 1 & 0 & 0 & 2 & & & & & \\ 1 & 0 & 0 & 0 & 1 & & & & \\ 1 & 2 & 2 & 2 & 0 & 4 & & & \\ 1 & 2 & 0 & 0 & 3 & 0 & 6 & & \\ 1 & 0 & 2 & 0 & 1 & 0 & 0 & 2 & \\ 1 & 0 & 0 & 2 & 1 & 0 & 0 & 0 & 2 \\ 1 & 2 & 2 & 2 & 3 & 4 & 6 & 6 & 6 & 12 \end{bmatrix}$$

We now consider some interesting results, such as the classical Pólya's enumeration theorem, which are specializations of the generalized Burnside's lemmas (LEMMA 4.15 and COROLLARY 4.16). These results gain computational advantage by restricting the group action A: S and the action matrix M or "weight function" W of LEMMA 4.15 or COROLLARY 4.16. The set S will be a set of functions, R^D, and the group A will act on these functions by permuting the elements of D. The cycle structure of permutations of D will play a critical role here, so we first review some ideas already familiar to the reader in order to collect together the necessary terminology.

Let D and R be sets. A function f from D to R is a rule that assigns to each element of D an element of R. Each element of D must be assigned something in R. The set D is called the *domain* of the function and the set B is called the *range*. The set {f(x): x ∈ D} is called the *image* of f.

4.34 EXAMPLE OF FUNCTIONS f: D → R.

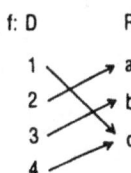

Here, the *domain* of the function is D, the *range* is R, the *image* is {a,c,d}. The arrows indicate the element of R that is being assigned to a particular element of D.

This function is *surjective* or *onto* because its image is equal to its range. To be precise, if y ∈ R, then there exists x ∈ D such that f(x) = y.

Figure 4.34, cont.

19

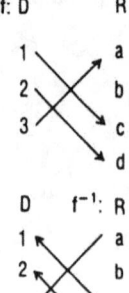

This function is *injective* or one-to-one because each element of D is assigned to a different element of R. In other words, if x,y ∈ D and x ≠ y, then f(x) ≠ f(y).

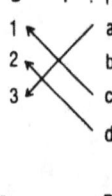

Every injective function has an "inverse" function denoted by f^{-1}. The domain of the inverse is the image of f. Clearly, f^{-1} is also injective. The function shown here is the inverse of the preceding example.

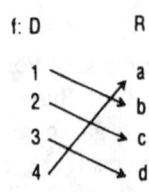

A function that is both injective and surjective is called bijective. The image of a bijective function is its range. In this case, f^{-1} is also a bijection. If f is a bijection, then D and R must have the same size (if they are finite). For finite sets, there exists a bijection f: D → R if and only if D and R have the same size (or "cardinality").

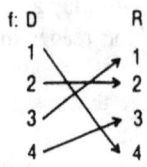

If f is a bijection and D equals R, then f is called a permutation of D. The set of all permutations of D is designated PER(D). When D = n̲ or ₍n₎, we have used S_n ≡ PER(D).

Figure 4.34

There are many ways of describing functions. Any such description must specify the domain, the range, and, for each element of the domain, the rule for computing the corresponding element of the range.

Here are some different ways of writing the function given by the first diagram in EXAMPLE 4.34: (1,a), (2,c), (3,a), (4,d) or $\begin{pmatrix} 1234 \\ acad \end{pmatrix}$ or "assign a to 1 and 3, c to 2, and d to 4." If D is infinite (all integers, the real numbers, etc.), then a tabulation of all values of the function is impossible. Instead, tables for some values are given if the function has a smooth graph (like sin and cos) or a rule for computing the corresponding value is given (like f(x) = x^2).

Of particular interest are ways of writing permutations of finite sets. The permutation of EXAMPLE 4.34 (the last diagram) can be written in three standard ways: $\begin{pmatrix} 1234 \\ 4213 \end{pmatrix}$ or 4213 or (143)(2). The first way is called "two-line notation," the second way is called "one-line notation," and the third way is called "cycle

20

notation." One-line notation is the same as two-line notation except that the domain values are just imagined to be there. The cycle notation is really different. The cycle f = (143) is read "1 goes to 4 (i.e., f(1) = 4), 4 goes to 3, 3 goes to 1." The cycle (2) is read "2 goes to 2." If the domain of the permutation is known, then cycles like (2) are omitted from the cycle notation. Thus, (143)(2) equals (143) *plus* the information that the domain is {1,2,3,4}.

If f and g are permutations then the "product" of f and g, written fg, is the *composition* of f and g as functions: fg(x) = f(g(x)). Suppose, for example, that f = (143) and g = (1562) are permutations of A = {1,2,3,4,5,6}. To "compose" f and g (i.e., to compute fg) we must compute for each integer x $\in$ A *first* g(x) and *then* f(g(x)). Take x = 5. Then g(5) = 6 and f(6) = 6. Thus, fg(5) = 6. For x = 2, g(2) = 1 and f(1) = 4. By going through the elements of the set A (in any order) one computes fg in two line notation fg = $\begin{pmatrix} 123456 \\ 541362 \end{pmatrix}$. In cycle notation fg = (156243). The reader should practice composing functions in both two-line notation and in cycle notation.

4.35 EXERCISE.

(1) Write down three pairs of permutations and compose them. Work both in two-line notation and in cycle notation.
(2) Write down five permutations of 1,2,3,4,5,6,7,8 in two-line notation and convert each one to cycle notation.

Every cycle can be written as a product of cycles of length two in many different ways. For example (1432) = (23)(14)(24) = (12)(13)(14) = (14)(42)(43) = (12)(13)(14)(43)(42)(14)(23)(14)(24). The reader should verify these expressions. Note that the number of cycles in the first three cases is three and in the last case is nine. A cycle of length two is called a *transposition*. It is a general result that whenever a cycle is written in different ways as a product of transpositions, then the number of transpositions in the different products have the same "parity." That is, both numbers are odd or both are even.

Two cycles are said to be disjoint if they have no entries in common. A permutation written such as (25174)(346) where the cycles are pair-wise disjoint is said to be written in "disjoint cycle" notation. By permuting the symbols cyclically in any cycle, that cycle is not changed *as a permutation*: (2517) = (5172) = (1725) = (7251). Also, in disjoint cycle notation the cycles themselves may be written down in any order ((2517)(346) or (346)(2517) for example). The *disjoint cycle notation* is unique up to these transformations. If any permutation is written in two different ways as a product of transpositions then the result mentioned above for single cycles is still true (the parity of the number of transpositions in each case is the same).

4.36 DEFINITION (Pólya action).

Suppose A acts on D, A: D, D finite. Let $S = R^D$, where both D and R are finite sets (i.e., S = all functions from D to R). Then we can define an action of A on R^D by

$$(af)(x) = f(a^{-1}x), \text{ for all } a \in A, f \in R^D, x \in D$$

Note that for a and b in A, $(ab)f(x) = f((ab)^{-1}x) = f(b^{-1}a^{-1}x) = a(bf(x))$ and hence $(ab)f = a(bf)$ so the rule of DEFINITION 4.36 does define an action of A on S. This action is called the *Pólya action* of A on S.

4.37 EXAMPLE OF PÓLYA ACTIONS.

(1) Let D = {1,2,3,4}, R = {g r t}, A = S_4. Consider τ = (1 2 3 4) $\in$ A. If
$$f = \begin{pmatrix} 1\,2\,3\,4 \\ g\,r\,g\,r \end{pmatrix}, \tau f = \begin{pmatrix} 1\,2\,3\,4 \\ r\,g\,r\,g \end{pmatrix}. \text{ If } f = \begin{pmatrix} 1\,2\,3\,4 \\ g\,r\,t\,r \end{pmatrix}, \tau f = \begin{pmatrix} 1\,2\,3\,4 \\ r\,g\,r\,t \end{pmatrix},$$
etc.

(2) $S = \{g,r\}^{\{1,2,3,4\}}$, A = $\{e,\tau,\tau^2,\tau^3\}$, where τ = (1 2 3 4). Note τ^2 = (1 3)(2 4), and S_{τ^2}, the functions unchanged by τ^2, are given by
$$S_{\tau^2} = \begin{pmatrix} 1\,3\,2\,4 \\ g\,g\,r\,r \end{pmatrix}, \begin{pmatrix} 1\,3\,2\,4 \\ r\,r\,g\,g \end{pmatrix}, \begin{pmatrix} 1\,3\,2\,4 \\ g\,g\,g\,g \end{pmatrix}, \begin{pmatrix} 1\,3\,2\,4 \\ r\,r\,r\,r \end{pmatrix}$$
$$\cong \begin{pmatrix} (1\,3)\,(2\,4) \\ g \quad\quad r \end{pmatrix}, \begin{pmatrix} (1\,3)\,(2\,4) \\ r \quad\quad g \end{pmatrix}, \begin{pmatrix} (1\,3)\,(2\,4) \\ g \quad\quad g \end{pmatrix}, \begin{pmatrix} (1\,3)\,(2\,4) \\ r \quad\quad r \end{pmatrix}.$$

In fact, for all a $\in$ A, $S_a = \{f: af = f\} \cong R^{\,cyc(a)}$ where cyc(a) is the set of all disjoint cycles of a (including cycles of length one).

The last observation of EXAMPLE 4.37(2) is extremely important to our present discussion: there is a natural correspondence between the functions unchanged by the action of a fixed permutation a $\in$ A and the set of *all* functions from cyc(a) to R, denoted by $R^{cyc(a)}$. This correspondence is shown in EXAMPLE 4.37(2) and extends in an obvious way to the general case.

In our discussion of the generalized Burnside lemma previously mentioned, we adopted the point of view that the objects of the sets themselves could be viewed as "variables" in polynomials. To adopt that point of view here would be to treat the functions f $\in$ R^D as variables. This is fine, and was done in EXAMPLE 4.18 where D was the vertices of the triangle and R = {g,r}. A natural variation of this idea in our present discussion is to think of the elements of D and R as variables. This can be done no matter what the elements of D and R are. In the following examples, we shall want to be free to choose D and R to suit the problem at hand. If however, D = {1,2,3} and we think of the objects of D as "variables" rather than integers, then one might confuse the

22

polynomial $2^2 + 2 + 3$ with an arithmetic statement about integers. If, on the other hand, we set $D = \{x_1, x_2, x_3\}$, then the same polynomial becomes $x_2^2 + x_2 + x_3$ and there is no confusion. When we wish to avoid such confusions, we shall choose $D = \{x_1, x_2, \ldots, x_d\}$ and $R = \{y_1, y_2, \ldots, y_r\}$ as sets of variables. Otherwise, we shall choose D and R for conceptual or notational convenience.

4.38 DEFINITION.

Let $D = \{x_1, x_2, \ldots, x_d\}$ and $R = \{y_1, y_2, \ldots, y_r\}$. Given $f \in R^D$, define

$$O(f) = \prod_{j=1}^{r} y_j^{|f^{-1}(y_j)|} = \prod_{i=1}^{d} f(x_i).$$

We call the reader's attention to the difference between $O(f)$ of DEFINITION 4.38 and $O(x)$ of EXAMPLE 4.18. The latter is an "Orbit selection" function and hence has different values on different orbits. $O(f)$ may equal $O(h)$ even though f and h are in different orbits of $A: R^D$ (see EXAMPLE 4.39(2)).

4.39 EXAMPLES OF DEFINITION 4.38.

(1) If $f = \begin{pmatrix} x_1 & x_2 & x_3 & x_4 \\ y_1 & y_2 & y_1 & y_2 \end{pmatrix}$, $O(f) = \prod_{i=1}^{4} f(x_i) = y_1 y_2 y_1 y_2 = y_1^2 y_2^2$.

(2) Note $O(S_a) = \sum_{f \in S_a} O(f) = \sum_{f \in S_a} \prod_{i=1}^{d} f(x_i)$. But since $S_a \cong R^{cyc(a)}$ the sum

$\sum_{f \in S_a} \prod_{i=1}^{d} f(x_i)$ is equal to $\sum_{g \in R^{cyc(a)}} \prod_{c \in cyc(a)} (g(c))^{|c|}$ where $g(c)$ is the value of f

at any element c of $cyc(a)$ and $|c|$ is the length (number of elements) of

the cycle c. For instance, if $f = \begin{pmatrix} 1 & 2 & 3 & 4 \\ a & b & a & b \end{pmatrix}$; $O(f) = \prod_{i=1}^{4} f(i) = a^2 b^2$;

$f \in S_{\tau^2}$ where $\tau^2 = (1\ 3)(2\ 4)$. But $f \cong g = \begin{pmatrix} (1\ 3)(2\ 4) \\ a \qquad b \end{pmatrix} \in R^{cyc(\tau^2)}$, and

$\prod_{c \in cyc(\tau^2)} (g(c))^{|c|} = a^{|(1\ 3)|} b^{|(2\ 4)|} = a^2 b^2 = O(f)$. Note that if $A = \{e, \tau, \tau^2, \tau^3\}$

is the cyclic group of order 4, then $h = \begin{pmatrix} 1 & 2 & 3 & 4 \\ a & b & b & a \end{pmatrix}$ also has $O(h) = a^2 b^2$,

but f and h are in different orbits of $A: R^D$. Observe that with $R = \{y_1, y_2, \ldots, y_r\}$, $\sum_{g \in R^{cyc(a)}} \prod_{c \in cyc(a)} (g(c))^{|c|} = \prod_{c \in cyc(a)} \sum_{j=1}^{r} y_j^{|c|}$ by interchanging product and sum. Therefore

$$O(S_a) = \sum_{f \in S_a} O(f) = \sum_{f \in S_a} \prod_{i=1}^{d} f(x_i)$$

23

$$= \sum_{g \in R^{cyc(a)}} \prod_{c \in cyc(a)} (g(c))^{|c|} = \prod_{c \in cyc(a)} \sum_{j=1}^{r} y_j^{|c|} .$$

Thus we have proved the basic PÓLYA ACTION IDENTITY 4.40.

4.40 PÓLYA ACTION IDENTITY AND AN EXAMPLE.

$$\sum_{f \in S_a} \prod_{i=1}^{d} f(x_i) = \prod_{c \in cyc(a)} \sum_{j=1}^{r} y_j^{|c|}$$

where $S_a = \{f: af = f\}$.

Let $R = \{y_1, y_2\}$, $a = (1\ 2)(3\ 4\ 5)$. Then

$$R^{cyc(a)} = \left(\begin{matrix} (1\ 2)(3\ 4\ 5) \\ y_1 \quad y_1 \end{matrix} \right), \left(\begin{matrix} (1\ 2)(3\ 4\ 5) \\ y_1 \quad y_2 \end{matrix} \right), \left(\begin{matrix} (1\ 2)(3\ 4\ 5) \\ y_2 \quad y_1 \end{matrix} \right), \left(\begin{matrix} (1\ 2)(3\ 4\ 5) \\ y_2 \quad y_2 \end{matrix} \right).$$

So

$$\sum_{g \in R^{cyc(a)}} \prod_{c \in cyc(a)} (g(c))^{|c|} = y_1^2 y_1^3 + y_1^2 y_2^3 + y_2^2 y_1^3 + y_2^2 y_2^3.$$

But

$$\prod_{c \in cyc(a)} \sum_{j=1}^{r} y_j^{|c|} = \left(\sum_{j=1}^{2} y_j^{|(3\ 4\ 5)|} \right) \left(\sum_{j=1}^{2} y_j^{|(1.2)|} \right)$$

$$= (y_1^3 + y_2^3)(y_1^2 + y_2^2)$$

$$= y_1^2 y_1^3 + y_1^2 y_2^3 + y_2^2 y_1^3 + y_2^2 y_2^3$$

$$= \sum_{g \in R^{cyc(a)}} \prod_{c \in cyc(a)} (g(c))^{|c|} .$$

Using COROLLARY 4.16 and IDENTITY 4.40, we can now easily prove Pólya's enumeration theorem.

4.41 THEOREM (Pólya's theorem).

Let $D = \{x_1, x_2, \ldots, x_d\}$ and let $R = \{y_1, y_2, \ldots, y_r\}$. Let the group A act on D and hence on $S = R^D$ by the Pólya action (DEFINITION 4.40). Then

$$\sum_{f \in \Delta} \left(\prod_{i=1}^{d} f(x_i) \right) = 1/|A| \sum_{a \in A} \prod_{c \in cyc(a)} \sum_{j=1}^{r} y_j^{|c|}$$

Proof. Using IDENTITY 4.40, this result is a special case of the weighted Burnside lemma COROLLARY 4.16(1). In COROLLARY 4.16(1), set $s = f$, $W(s) = \prod_{i=1}^{d} f(x_i)$. Then $\sum_{s \in S_a} W(s) = \sum_{f \in S_a} \prod_{i=1}^{d} f(x_i) = \prod_{c \in cyc(a)} \left(\sum_{j=1}^{r} y_j^{|c|} \right)$ by IDEN-TITY 4.40.

24

Another standard formulation of THEOREM 4.41 is in terms of the polynomial $P_A(z_1, \ldots, z_d)$ associated with the action A: D called the *cycle index polynomial*.

4.42 DEFINITION.

For each permutation $a \in A$, let $\nu(a,i)$ denote the number of cycles in a of length i (as a permutation of D). The vector $(\nu(a,1), \ldots, \nu(a,d))$ is called the *type* of a. The polynomial

$$P_A(z_1, \ldots, z_d) = \frac{1}{|A|} \sum_{a \in A} z_1^{\nu(a,1)} z_2^{\nu(a,2)} \cdots z_d^{\nu(a,d)}$$

is called the *cycle index polynomial of A acting on D*.

Using DEFINITION 4.42, we have the obvious rephrasing of Pólya's theorem as IDENTITY 4.43.

4.43 CYCLE INDEX VERSION OF PÓLYA'S THEOREM.

$$\sum_{f \in \Delta} \prod_{i=1}^{d} f(x_i) = P_A\left(\sum_{j=1}^{r} y_j, \sum_{j=1}^{r} y_j^2, \ldots, \sum_{j=1}^{r} y_j^d \right).$$

The main feature of Pólya's theorem as distinct from the general Burnside lemma (from which Pólya's theorem is obtained) is that it allows information about the range R to be included as a variable or "parameter" in the formulas in a particularly nice way. We now consider some examples.

The symmetries of the cube under rotations only (no reflections) is shown in FIGURE 4.44. Imagine the faces of the cube labeled 1 through 6. Looking at the cube as in FIGURE 4.44(b), we label the front face 1, the top face 2, the back face 3, the bottom face 4, the right-side face 5, and the left-side face 6. The permutation (1 2 3 4)(5)(6) is interpreted as "1 is sent to where 2 used to be, 2 is sent to where 3 used to be, etc.," just as in the case of the symmetries of the square in FIGURE 4.2. FIGURE 4.44 shows how the cycle index polynomial is computed. The number of distinct cubes with r colors is gotten by setting $y_j = 1$, $j = 1, \ldots, r$, in IDENTITY 4.43. The reader is asked to explore some additional ideas related to Pólya's theorem in EXERCISE 4.45.

4.44 ROTATION GROUP OF CUBE ACTING ON FACES.

```
            4
            3
          6 2 5
            1 ← Front face
```

(a) identity = e   type = (6, 0, 0, 0, 0, 0)
                         (see DEF. 4.42)   Figure 4.44 (cont.)

(b) 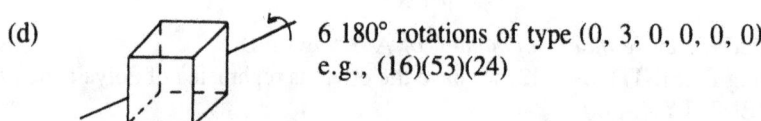 6 90° rotations of type = (2, 0, 0, 1, 0, 0)
e.g., (1 2 3 4)(5)(6)

(c) 3 180° rotations of type (2, 2, 0, 0, 0, 0)
e.g., (13)(24)(5)(6)

(d) 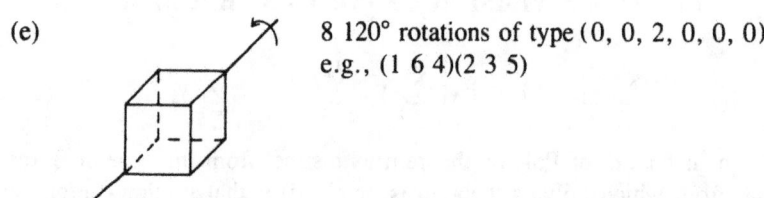 6 180° rotations of type (0, 3, 0, 0, 0, 0)
e.g., (16)(53)(24)

(e) 8 120° rotations of type (0, 0, 2, 0, 0, 0)
e.g., (1 6 4)(2 3 5)

Figure 4.44
(continued)

$$P_A(z_1, z_2, z_3, z_4, z_5, z_6) = \frac{1}{24}(z_1^6 + 6z_1^2 z_4 + 3z_1^2 z_2^2 + 6z_2^3 + 8z_3^2)$$

NUMBER OF DISTINCT CUBES
WITH r COLORS (FACES) $= \frac{1}{24}(r^6 + 12r^3 + 3r^4 + 8r^2).$

4.45 EXERCISE.

(1) The group of rotations of the cube, considered in FIGURE 4.44, also acts on the edges and the vertices of the cube in an analogous fashion. What are the cycle index polynomials for these actions?

(2) The full group of symmetries of the cube includes both rotations and reflections and has 48 elements. What is the cycle index polynomial of this group?

(3) The rotational group of symmetries of the regular polygon with d vertices is the cyclic group Z_d generated by the cycle $(1,2,3,\ldots,d)$ (if we think of this group acting on vertices). Show that the cycle index polynomial of this action is given by $P_{Z_d}(z_1,\ldots,z_d) = \frac{1}{d} \sum_{i|d} \varphi(i) z_i^{d/i}$ where the sum is over all

divisors of d and $\varphi(i)$ is the number of integers less than or equal to i and relatively prime to i (Euler's φ-function).

(4) The group of symmetries of the regular polygon with d vertices allowing reflections has 2d elements and is called the *dihedral group* D_{2d}. We have considered certain aspects of this group in the previous material (d = 4 in FIGURE 4.2, d = 3 in FIGURE 4.11, d = 6 in EXERCISE 4.32(3)). Prove that the cycle index polynomial $P_{D_{2d}}(z_1, \ldots, z_d) = \frac{1}{2}(P_{Z_d}(z_1, \ldots, z_d) + \frac{1}{2}(z_2^{d/2} + z_1^2 z_2^{(d-2)/2}))$ if d is even and $P_{D_{2d}}(z_1, \ldots, z_d) = \frac{1}{2}(P_{Z_d}(z_1, \ldots, z_d) + z_1 z_2^{(d-1)/2})$ if d is odd.

(5) Using IDENTITY 4.22, extend Pólya's theorem to include group characters. Illustrate your theorem with the rotational group of symmetries of the polygon with d vertices (i.e., the cyclic group Z_d). Recall EXAMPLES 4.23 and 4.24. Prove White's theorem, the Pólya theorem analog of Lemma 4.29.

As our final example of a cycle index polynomial, we consider the cycle index polynomial of the symmetric group on $\underline{d} = \{1, \ldots, d\}$. The result is of theoretical rather than practical importance as the index of summation is over a set whose size grows exponentially with d. We denote by S_d the group of all permutations (symmetric group) of $\underline{d}$. For $a \in S_d$ we define the vector $(\nu(a,1), \nu(a,2), \ldots, \nu(a,d))$ to be the *type* of a (the $\nu(a,i)$ are as in DEFINITION 4.42). For example, if $a = (1\ 2\ 3)(4\ 5)(6\ 7)(8)(9)$, then type (a) = (2, 2, 1, 0, \ldots, 0). So

$$P_{S_d} = 1/d! \sum_{\substack{(\alpha_1, \ldots, \alpha_d) \\ \alpha_1 + 2\alpha_2 + \ldots + d\alpha_d = d}} |\{a: \text{type } (a) = (\alpha_1, \ldots, \alpha_d)\}| z_1^{\alpha_1} \ldots z_d^{\alpha_2}.$$

Now let us try to determine, given type $(\alpha_1, \alpha_2, \ldots, \alpha_d)$, how many permutations have this type?

If a has type $(\alpha_1, \ldots, \alpha_d)$, then the "form" of the cycle decomposition of a is

$$a = \underbrace{(\square)(\square) \ldots (\square)}_{\alpha_1 \text{ 1-cycles}}\underbrace{(\square\square)(\square\square) \ldots (\square\square)}_{\alpha_2 \text{ 2-cycles}}$$

$$\ldots \underbrace{(\square \ldots \square) \ldots (\square \ldots \square)}_{\alpha_k \text{ k-cycles}}$$

$$\vdots \underbrace{(\square \ldots \square) \ldots (\square \ldots \square)}_{\alpha_d \text{ d-cycles } (\alpha_d = 0 \text{ or } 1)}$$

There are d! different ways of putting integers in the above boxes. Each placement may be regarded as a permutation of type $(\alpha_1, \ldots, \alpha_d)$. But not all of these will give a *different* permutation, for instance $(5)(4)(3\ 2\ 1) = (4)(5)(2\ 1\ 3)$, etc. It turns out that by dividing by the number of such duplications we obtain

$$|\{a: \text{type }(a) = (\alpha_1, \alpha_2, \ldots, \alpha_d)\}| = \frac{d!}{\alpha_1! \alpha_2! \ldots \alpha_d! 1^{\alpha_1} 2^{\alpha_2} \ldots d^{\alpha_d}}.$$

Thus, we obtain IDENTITY 4.46.

4.46 CYCLE INDEX POLYNOMIAL OF SYMMETRIC GROUP.

$$P_{S_d}(z_1, \ldots, z_d) = \frac{1}{d!} \sum_{\substack{(\alpha_1, \ldots, \alpha_d) \\ \alpha_1 + 2\alpha_2 + \ldots + d\alpha_d = d}} \frac{d!}{\displaystyle\prod_{i=1}^{d} \alpha_i! i^{\alpha_i}} z_1^{\alpha_1} \ldots z_d^{\alpha_d}.$$

4.47 EXERCISE.

(1) Let $P_{S_d}(z_1, \ldots, z_d)$ be the cycle index polynomial of the symmetric group S_d. Prove that

$$P_{S_d}(z_1, \ldots, z_d) = \frac{1}{d} \sum_{k=1}^{d} z_k P_{S_{d-k}}(z_1, \ldots, z_{d-k}).$$

For small d this provides a useful recursion for calculating P_{S_d}. Try it for d = 1,2,3,4.

(2) Prove that $P_{S_d}(z_1, \ldots, z_d)$ is the coefficient of u^d in the development of

$$\exp\left(uz_1 + \frac{u^2 z_2}{2} + \frac{u^3 z_3}{3} + \ldots \right)$$

as a power series in u.

(3) Consider Pólya's theorem (IDENTITY 4.43) applied to the group of rotational symmetries of the cube (FIGURE 4.44). Suppose $R = \{y_1, y_2, y_3\}$ has three elements ("colors"). Let $y_1 = u$, $y_2 = y_3 = 1$. What is the interpretation of the left-hand side of IDENTITY 4.43? Such a "change of variables" for the values of R is called an "assignment of Pólya weights." Try to think up some imaginative assignments of Pólya weights for the various Pólya actions we have constructed thus far (cube groups, cyclic groups, dihedral groups, symmetric groups).

There are two basic techniques for extending the scope of Pólya's theorem. One may attempt to systematically enlarge the class of group actions A: D for which reasonable formulas are known for the cycle index polynomial, or one may enlarge the class of group actions analogous to the Pólya action for which a similar result may be obtained. We shall conclude our discussion of Pólya's theorem with some examples of each approach.

Let A and B be groups and P and Q sets such that A: P and B: Q. The set of all functions from P to B, B^P, is also a group where, for φ and ψ in B^P, i in P, $\varphi\psi(i) = \varphi(i)\psi(i)$. This is called the "pointwise product of φ and ψ" for obvious reasons. Each pair (a,φ), $a \in A$ and $\varphi \in B^P$, may be regarded as a permutation of the set of all pairs $P \times Q = \{(i,j): i \in P, j \in Q\}$ by defining $(a,\varphi)(i,j) = (ai, \varphi(i)j)$. With this definition we see that composition of permutations of this form goes as follows: $(a_1,\varphi_1)(a_2,\varphi_2)(i,j) = (a_1,\varphi_1)(a_2i,\varphi_2(i)j)$ $= (a_1a_2i,\varphi_1(a_2i)\varphi_2(i)j)$. Thus, the permutations of this type are closed under composition (i.e., multiplication of permutations) according to the rule $(a_1,\varphi_1)(a_2,\varphi_2)$ $= (a_1a_2,(\varphi_1a_2)\varphi_2)$. The identity permutation is of this form: (e,ε) where e is the identity in A and ε is the function that maps each element of P to the identity of B. If (a,φ) is a permutation, then (a^{-1},ψ) is its inverse where ψ is the inverse of φa^{-1} in B^P. Thus, the set of all permutations of the form (a,φ) with $a \in A$ and $\varphi \in B^P$ forms a subgroup of the group of all permutations PER(PxQ). This permutation group will be called the *wreath product* of A: P and B: Q and will be denoted by A[B]. The reader familiar with a little group theory will note that the wreath product is an action of a semidirect product of A and B^P. The group B^P is a normal subgroup of this product. We regard A and B as subgroups of A[B] by the obvious identifications.

4.48   WREATH PRODUCT OF S_3 AND S_2 ACTING ON $\underline{3} \times \underline{2}$.

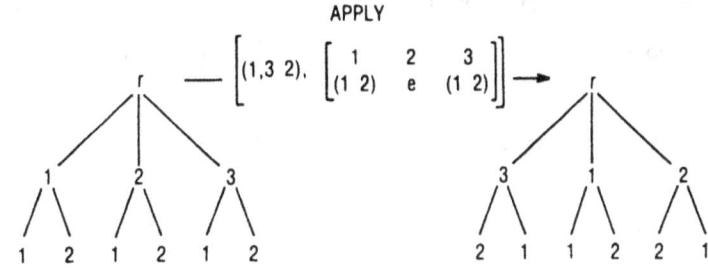

Figure 4.48

29

An example of the product of elements in a wreath product is given in FIGURE 4.48. There, we consider the wreath product of S_3 and S_2 acting on $\underline{3} \times \underline{2}$. The set $\underline{3} \times \underline{2}$ is represented by the standard ordered tree diagram (see FIGURES 1.42 and 3.6). In FIGURE 4.48, we compose two elements of $S_3[S_2]$. The composition is $(a,\varphi)(b,\psi)$ where $a = (1\ 3)$, $b = (1\ 3\ 2)$, $\varphi = \begin{pmatrix} 1 & 2 & 3 \\ e & e & (1\ 2) \end{pmatrix}$ and

$\psi = \begin{pmatrix} 1 & 2 & 3 \\ (1\ 2) & e & (1\ 2) \end{pmatrix}$. The reader should check carefully the computation of FIGURE 4.48 against the definitions of the previous paragraph. In general, a wreath product $A[B]$ acting on $P \times Q$ has $|A||B|^{|P|}$ elements. Thus, $S_3[S_2]$ has 48 elements and acts on a set, $\underline{3} \times \underline{2}$, which has six elements. The reader who has worked EXERCISE 4.45(2) will be struck by the similarities of this group with the full symmetry group of the cube (48 elements) acting on the faces of the cube (six elements). In fact, the wreath product $S_3[S_2]$ is exactly the group of symmetries of the cube acting on the faces of the cube (up to certain natural identifications). More generally, $S_d[S_2]$ is the full symmetry group of the d-dimensional cube acting on faces. The basic idea is shown in FIGURE 4.49. The set P is identified with the three oriented axes as shown by the labels 1, 2, and 3. The endpoints of each axis are identified with the labels 1 and 2 corresponding to the set Q. One can easily verify how the group $S_3[S_2]$ relates to the symmetries of the cube. For example, reflection through the plane passing through the left-most and right-most vertical edges of the diagram of FIGURE 4.49 corresponds to the element $\left((2\ 3), \begin{pmatrix} 1 & 2 & 3 \\ e & e & e \end{pmatrix} \right)$ of the wreath product.

4.49 FULL SYMMETRY GROUP OF CUBE AS A WREATH PRODUCT.

A = $S_3[S_2]$ with P = $\underline{3}$ and Q = $\underline{2}$ has order 48

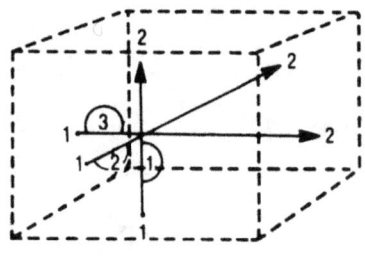

Figure 4.49

In FIGURE 4.50 we see a square whose vertices are labeled with labeled triangles. The vertices of the triangles are labeled with symbols g,r. Imagine that the square can be rotated 0°, 90°, 180°, or 270° and at the same time the triangles can be rotated independently of the square and of each other by 0°,

30

120°, or 240°. This is intuitively what is meant by the action of the wreath product $C_4[C_3]$ on the structures such as that of FIGURE 4.50. If we think of the vertices of the square as $\{1,2,3,4\} = \underline{4}$ and the vertices of the triangle as $\{1,2,3\} = \underline{3}$, then the twelve vertices of FIGURE 4.50 can be identified with $\underline{4} \times \underline{3}$. The transformations of FIGURE 4.50 just described represents the Pólya action of $C_4[C_3]$ on $\{g,r\}^{\underline{4} \times \underline{3}}$.

Again referring to FIGURE 4.50, the set $\Delta_t(3)$ is a system of orbit representatives for the Pólya action of C_3 on $\{g^t,r^t\}^{\underline{3}}$, t a positive integer. By Pólya's theorem (IDENTITY 4.43) we have

$$\sum_{f \in \Delta_t(3)} \prod_{i=1}^{3} f(i) = g^{3t} + g^{2t}r^t + g^t r^{2t} + r^{3t} \equiv \eta_t$$

$$= P_{C_3}(g^t + r^t, g^{2t} + r^{2t}, g^{3t} + r^{3t}) .$$

We may also consider the Pólya action of C_4 on $(\Delta_1(3))^{\underline{4}}$. That is, we imagine the vertices of the square of FIGURE 4.50 being labeled in all possible ways with elements of $\Delta_1(3)$. The structure shown in FIGURE 4.50 is one such labeling. It is intuitively obvious that an orbit representative system for $C_4[C_3]$ acting on $(\Delta_1(3))^{\underline{4}}$ is also an orbit representative system for $C_4[C_3]$ acting on $\{g,r\}^{\underline{4} \times \underline{3}}$. If in each case we take the Pólya weights (see EXERCISE 4.47(3)) to be the product of all labels g or r that occur in the structure we consequently obtain the wreath product identity for $C_4[C_3]$ (IDENTITY 4.51).

4.50 ACTION OF WREATH PRODUCT $C_4[C_3]$.

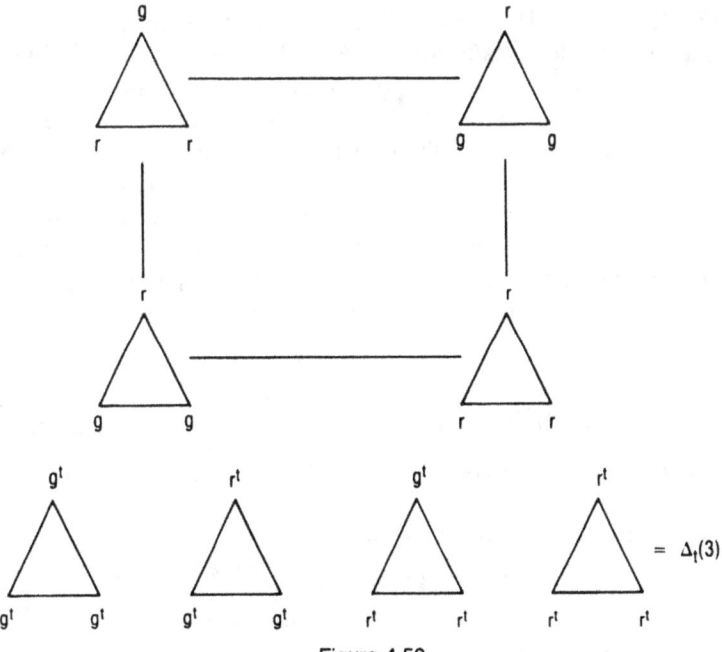

Figure 4.50

31

4.51 WREATH PRODUCT IDENTITY FOR $C_4[C_3]$.

$$P_{C_4[C_3]}(g + r, \ldots, g^t + r^t, \ldots, g^{12} + r^{12}) = P_{C_4}(\eta_1, \eta_2, \eta_3, \eta_4)$$

where

$$\eta_t = g^{3t} + g^{2t}r^t + g^t r^{2t} + r^{3t} = P_{C_3}(g^t + r^t, g^{2t} + r^{2t}, g^{3t} + r^{3t}).$$

This is the sense in which a cycle index polynomial of a wreath product is a composition of cycle index polynomials. Instead of just having $R = \{g, r\}$ we might have $R = \{y_1, \ldots, y_r\}$ and we might have the general case of $A[B]$ acting on $P \times Q$ ($|P| = p$, $|Q| = q$). Then IDENTITY 4.51 becomes IDENTITY 4.52.

4.52 WREATH PRODUCT IDENTITY FOR THE GENERAL CASE.

$$P_{A[B]}\left(\sum_{j=1}^{r} y_j, \ldots, \sum_{j=1}^{r} y_j^{pq}\right) = P_A(\eta_1, \eta_2, \ldots, \eta_p)$$

where

$$\eta_t = P_B\left(\sum_{j=1}^{r} y_j^t, \sum_{j=1}^{r} y_j^{2t}, \ldots, \sum_{j=1}^{r} y_j^{qt}\right).$$

By making the change of variable $z_k = \sum_{j=1}^{r} y_j^k$ for $k = 1, \ldots, pq$ we obtain THEOREM 4.53. There is a slight technical difficulty in asserting that this change of variable actually transforms IDENTITY 4.52 into THEOREM 4.53 because in one case we have a polynomial in $y_1, \ldots, y_r$, and in the other case a polynomial in $z_1, \ldots, z_{pq}$. But, as we are free to choose r as large as we please it is hard to imagine the result not being true. In fact, $r = pq$ is good enough by a standard theorem in algebra about symmetric functions.

We now give a completely different and independent proof of THEOREM 4.53. In this proof, we analyze directly the cycle structure of a wreath product. In spite of some notational complexity, the idea of the proof is beautifully simple, involving only a product-sum interchange (see STEP 4).

4.53 THEOREM.

Let A: P, B: Q with $|P| = p$, $|Q| = q$. Let $A[B]$ be the wreath product. Then

$$P_{A[B]}(z_1, \ldots, z_{pq}) = P_A(P_B(z_1, \ldots, z_q), P_B(z_2, \ldots, z_{2q}), \ldots, P_B(z_p, \ldots, z_{pq})) .$$

Proof. The theorem asserts that to construct $P_{A[B]}(z_1, \ldots, z_{pq})$, one first constructs $P_A(z_1, \ldots, z_p)$ and then, for each z_t, substitutes $P_B(z_t, z_{t2}, \ldots, z_{tq})$. By definition, $|A[B]| \, P_{A[B]}(z_1, \ldots, z_{pq}) = \sum_{(a, \varphi) \in A[B]} z_1^{\nu(a, \varphi, 1)} \ldots z_{pq}^{\nu(a, \varphi, pq)}$ where $\nu(a, \varphi, t)$ is the number of cycles of (a, φ) of length t.

32

STEP 1. Fix $(a,\varphi) \in A[B]$ and $(i,j) \in P \times Q$. Suppose that i is in a cycle of a of length 3 (as A acts on P). The cycle of (a,φ) containing (i,j) can be written

$$((i,j),(ai,\varphi(i)j),(a^2i,\varphi(ai)\varphi(i)j),(i,\varphi(a^2i)\varphi(ai)\varphi(i)j)...) .$$

Let $b = \varphi(a^2i)\varphi(ai)\varphi(i) \in B$ and suppose that j is in a cycle of length h of b (as B acts on Q). Thus, the cycle of j is $(j,bj,...,b^{h-1}j)$. Then the cycle of (a,φ) containing (i,j) can be written $((i,j),...,(i,bj),...,(i,b^{h-1}j),...)$ and has length 3h. If i is contained in a cycle of length g of a then we set $b = \varphi(a^{g-1}i)...\varphi(i)$ and, if j is in a cycle of b of length h we find that (i,j) is in a cycle of (a,φ) of length gh.

STEP 2. As a result of STEP 1, for each cycle $c = (i,ai,...,a^{g-1}i)$ of $a \in A$, there is a permutation $b_c = \varphi(a^{g-1}i)...\varphi(i)$. If $cyc(a)$ denotes the cycles of a, the correspondence $c \rightarrow b_c$ is a function from $cyc(a)$ to B. Let $|c|$ denote the length of c. Each permutation (a,φ) contributes the term

$$\prod_{c \in cyc(a)} \prod_{c' \in cyc(b_c)} z_{|c|\,|c'|} \quad \text{to} \quad P_{A[B]}$$

STEP 3. We now count the number of different functions φ such that (a,φ) produces the same correspondence $c \rightarrow b_c$ and hence the same terms of STEP 2. Let $b_c = \varphi(a^{g-1}i)...\varphi(ai)\varphi(i)$. The value of φ may be specified arbitrarily on the values $a^{g-1}i,...,ai$. We may then choose $\varphi(i)$ so that the product is b_c. Thus, we have $|B|^{|c|-1}$ such choices for each cycle c and hence each map $c \rightarrow b_c$ contributes

$$\prod_{c \in cyc(a)} |B|^{|c|-1} \prod_{c \in cyc(a)} \prod_{c' \in cyc(b_c)} z_{|c|\,|c'|} \quad \text{to} \quad P_{A[B]}$$

STEP 4. This is the key step and involves a product-sum interchange. From STEP 3 we now see that each fixed $a \in A$ contributes (to $P_{A[B]}$)

$$\prod_{c \in cyc(a)} |B|^{|c|-1} \sum_{c \rightarrow b_c} \prod_{c} \prod_{c'} z_{|c|\,|c'|}$$

where the sum is over all maps $c \rightarrow b_c$ from $cyc(a)$ to B. By the standard rule for interchanging sums and products this becomes

$$\prod_{c \in cyc(a)} |B|^{|c|-1} \prod_{c \in cyc(a)} \sum_{b \in B} \prod_{c' \in cyc(b)} z_{|c|\,|c'|} \; .$$

STEP 5. (Last step!) This step is nothing more than a change of notation. As usual, let $\nu(a,k)$ denote the number of cycles of $a \in A$ of length k. Note that

$$\prod_{c \in cyc(a)} |B|^{|c|-1} = |B|^{|P|} \prod_{c \in cyc(a)} \frac{1}{|B|} .$$ Thus, the last expression of STEP 4 becomes

$$|B|^{|P|} \prod_{k=1}^{P} \left(\frac{1}{|B|} \sum_{b \in B} z_{k1}^{\nu(b,1)} ... z_{kq}^{\nu(b,q)} \right)^{\nu(a,k)} .$$

Summing over all a $\in$ A we thus obtain the sum over all $(a,\varphi) \in$ A[B] of terms $z_1^{\nu(a,\varphi,1)} \ldots z_{pq}^{\nu(a,\varphi,pq)}$ as $|A| \, |B|^{|P|} \, P_A(P_B(z_1,\ldots,z_q),\ldots, P_B(z_p,\ldots,z_{pq}))$. Dividing by $|A| \, |B|^{|P|} = |A[B]|$ gives the result.

4.54 EXERCISE.

(1) Let $\mathscr{P}(n)$ denote the set of all polynomials $p(x_1,\ldots,x_n)$ in n variables. The symmetric group S_n acts on $\mathscr{P}(n)$ by the rule, $\sigma \in S_n$, $\sigma p(x_1,\ldots,x_n) = p(x_{\sigma^{-1}1},\ldots,x_{\sigma^{-1}n})$. Let $(S_n)_p$ denote the stability subgroup at p (i.e., all σ such that $\sigma p = p$). For $p(x_1,\ldots,x_9) = x_1x_2x_3 + x_4x_5x_6 + x_7x_8x_9$, what is $(S_9)_p$? What is the size of the orbit containing p as S_9 acts on $\mathscr{P}(9)$?

(2) We consider structures such as those shown in FIGURE 4.50. Instead of just two labels $\{g,r\}$, consider labels $\{y_1,y_2,\ldots,y_r\}$ (now "r" is a parameter counting the number of labels, not a label). Give explicit polynomials in r that count the number of labeled structures up to the action of $C_4[C_3]$, $D_8[C_3]$, $C_4[D_6]$, and $D_8[D_6]$. Recall that D_8 is the dihedral group of all rotations and reflections of the square and D_6 is the group of all rotations and reflections of the triangle. Graph these polynomials and compare their behavior for large values of r.

(3) Let A and B be groups, A: P and B: Q. The *direct product* A $\times$ B of A and B is the group of ordered pairs (a,b) with a $\in$ A and b $\in$ B where multiplication is defined by the rule $(a,b)(a',b') = (aa',bb')$. A $\times$ B acts on P $\times$ Q in the obvious way: $(a,b)(i,j) = (ai,bj)$. Express the cycle index polynomial of A $\times$ B: P $\times$ Q in terms of the cycle index polynomials of A: P and B: Q.

(4) Give procedures for listing, ranking, and unranking $S_m[S_n]$.

(5) Construct a bijection between the orbits of the Pólya action of A[B] on $R^{P \times Q}$ and the Pólya action of A on $(\Delta(Q))^P$ where $\Delta(Q)$ is the list of orbits of the Pólya action of B on R^Q. Give careful proofs.

(6) Describe explicitly an orbit representative system for the Pólya action of $S_m[S_n]$ on $R^{\underline{m} \times \underline{n}}$ where $R = y_1,y_2,\ldots,y_r$ is a linearly ordered set. How would you linearly order, rank, and unrank this list? Hint: The nondecreasing functions are a list of orbit representatives for S_n: $R^{\underline{n}}$. Recalling EXERCISE 3.26 one can then use (5) above. Give explicit formulas for the number of orbits of the action of $S_m[S_2]$ on the faces of the m-cube with r possible labels. Note that Pólya's theorem is not needed for this exercise.

We now consider how one can develop analogous theorems to Pólya's theorem for other group actions. There is by now an extensive literature on the subject of Pólya theory or "pattern enumeration." We shall only give a brief description of some of the most basic ideas and refer the reader to the references at the end of the chapter for further study. Much of what we now describe is based on the extensive work of N. G. deBruijn on pattern enumeration.

34

4.55 DEFINITION.

Let A be a group such that A acts on a set P and a set Q. The *Cartesian action* of A on $P \times Q$ is defined by $a(i,j) = (ai,aj)$ where $a \in A$, $i \in P$, $j \in Q$.

The Cartesian action has many variations. First of all, if G and H are groups such that $G: P$ and $H: Q$, then we can set $A = G \times H$ to be the direct product of G and H. Then for (g,h) in A, define $(g,h)i = gi$ for all $i \in P$. Thus, A acts on P. Similarly, A acts on Q. In this case the Cartesian action of A on $P \times Q$ satisfies $(g,h)(i,j) = ((g,h)i,(g,h)j) = (gi,hj)$. Thus, the direct product action of EXERCISE 4.54(3) is a special case of the Cartesian action. Given any function $f \in Q^P$ we define the graph of f to be the set $GRAPH(f) = \{(i,j): i \in P, j \in Q$ such that $f(i) = j\}$, $GRAPH(f)$ is a subset of $P \times Q$. If A acts on $P \times Q$ by the Cartesian action (or any action for that matter) then A acts on the set of all subsets of $P \times Q$. In particular, for $a \in A$, $aGRAPH(f) = \{(ai,aj): f(i) = j\}$ $= \{(s,t): f(a^{-1}s) = a^{-1}t\} = GRAPH(afa^{-1})$. In other words, A acts on the set of functions Q^P by acting on their graphs. Thus, the action $f \to afa^{-1}$ on Q^P is a special case of the Cartesian action. If $A = G \times H$ is the direct product action, and (g,h) is in A, then we have $(g,h)f = hfg^{-1}$. Finally, if A acts as the identity on Q, then $f \to fa^{-1}$ is the Pólya action. We also should note that the Cartesian action on functions sends injective functions to injective functions and also sends surjective functions to surjective functions.

We return now to the notation used in connection with Pólya's theorem. Let $D = \{x_1, \ldots, x_d\}$ and let $R = \{y_1, \ldots, y_r\}$. Let R^D denote the set of all functions from D to R.

4.56 DEFINITION.

Let A be a group, $A: R$ and $A: D$. Let ν be the homomorphism from A to $PER(R^D)$ defined by $\nu(a)(f) = afa^{-1}$. Then ν defines an action of A on R^D that we call the *Cartesian action on functions*.

35

4.57 INTUITIVE IDEA OF LEMMA 4.58.

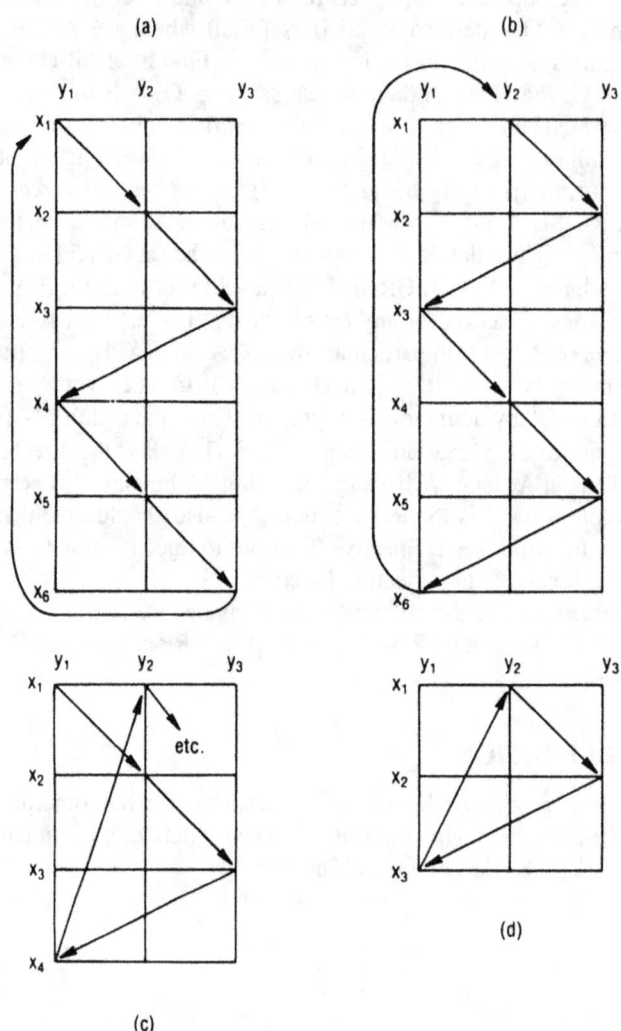

Figure 4.57

As just stated, the Cartesian action on functions has both the injective and surjective functions as invariant subsets of R^D. Just as in the case with Pólya's theorem, we shall apply the general Burnside's lemma (COROLLARY 4.16) to obtain an analog of THEOREM 4.41. As with these earlier results, the central problem is to describe the set S_a of all functions f such that $afa^{-1} = f$ for a fixed $a \in A$. Let $cyc_D(a)$ denote the set of cycles of the cycle decomposition of a acting on D and similarly define $cyc_R(a)$. We describe the intuitive idea involved in characterizing S_a in terms of the Cartesian action of A on GRAPH(f). Thus, we want to have aGRAPH(f) = GRAPH (f). This means that if $(x_i, y_j) \in$

GRAPH(f) then $(ax_i, ay_j) \in$ GRAPH(f). FIGURE 4.57 shows the sort of things that can happen.

Suppose we have an element $a \in A$ and suppose that $c = (x_1, . . ., x_6) \in$ $\text{cyc}_D(a)$. Suppose also that $c' = (y_1, y_2, y_3) \in \text{cyc}_R(a)$. FIGURE 4.57(a) shows the cycle of $a \in A$ acting on $D \times R$ containing (x_1, y_1). This cycle, indicated by the arrows in FIGURE 4.57(a), is $((x_1, y_1), (ax_1, ay_1), . . ., (a^5x_1, a^5y_1))$. It is evident that if $f \in S_a$ and hence aGRAPH(f) = GRAPH(f), then the condition $(x_1, y_1) \in$ GRAPH(f) forces all other points (ax_1, ay_1). . . in this cycle to be in GRAPH(f). In the same manner, if $(x_1, y_2) \in$ GRAPH(f) then all other points shown in FIGURE 4.57(b) must be in GRAPH(f). There is one other possibility, namely $(x_1, y_3) \in$ GRAPH(f). For both FIGURES 4.57(a) and (b) it is quite possible to have functions whose graphs are as shown. Thus, if $f \in S_a$ and $f(x_1) \in \{y_1, y_2, y_3\}$, then there are only $|c'| = 3$ possibilities for f restricted to c.

Now consider FIGURE 4.57(c). Here $c = (x_1, . . ., x_4)$ and $c' = (y_1, y_2, y_3)$. Again, suppose that $f \in S_a$ and $(x_1, y_1) \in$ GRAPH(f). For the same reasons, we must have all elements of the cycle $((x_1, y_1), (ax_1, ay_1), . . .)$ in GRAPH(f). But, from FIGURE 4.57(c) it is clear that this means that both (x_1, y_1) and (x_1, y_2) must be in GRAPH(f). This is impossible since f is a function. Comparing the situation in FIGURES 4.57(a) and (b) with that in FIGURE 4.57(c), we see that for $c \in \text{cyc}_D(a)$ and $c' \in \text{cyc}_R(a)$ and $x \in c$, it is possible to have $f(x) \in c'$, $f \in S_a$, *only if* the length of c' divides the length of c (we write $|c'| \, \big| \, |c|$). In the case where $|c'| \, \big| \, |c|$, there are exactly $|c'|$ values we can assign to $f(x)$. Once $f(x)$ is specified, then the value of f at all other points of c is determined. These observations obviously extend to the general case and clearly specify how one might construct all elements of S_a.

If we denote by R_*^D the injective maps from D to R, then, as we have observed above R_*^D is invariant under the Cartesian action of A. If f is an injective function and $f \in S_a$, then f must be of the type described in the previous paragraph. The fact that f is injective, however, rules out situations such as that shown in FIGURES 4.57(a) and (b) as these cannot be the graphs of an injective function.

It is easily seen that in addition to having $|c'| \, \big| \, |c|$, we must in fact have $|c'| = |c|$ for injective maps. But there is still another difference between the general case and injective map! If c and c' are as above and $x \in c$, $f(x) \in c'$ then f is an injection from c to c' (assume $f \in R_*^D$). In this case we say that f *associates* c' to c. If $|c| = k$ then there are, in general, other cycles of $a \in A$: D of length k. Let $\nu_D(a, k)$ denote the number of cycles of $a \in A$ of length k (as a permutation of D). Similarly, define $\nu_R(a, k)$. Clearly, no fixed c' can be associated by f to more then one cycle c if f is injective. In other words, associated with each injection, $f \in R_*^D$, and each integer k, there is an injection θ from the k-cycles of a in D to the k-cycles of a in R. The map θ is defined by $\theta(c) = c'$ if f associates c' to c. Let $\text{Im}(\theta)$ be the image of θ. Let $\text{cyc}_D(a, k)$ be the set of cycles of $a \in A$: D of length k and similarly define $\text{cyc}_R(a, k)$. The set of all injective maps θ from $\text{cyc}_D(a, k)$ to $\text{cyc}_R(a, k)$ will be denoted by $\mathscr{I}_k(a)$. LEMMA 4.58 summarizes the ideas of the previous two paragraphs and FIGURE 4.57 and is the extension of the Pólya action IDENTITY 4.40 to the Cartesian action of A on R^D and R_*^D (DEFINITION 4.56). Recall X of NOTATION 4.27.

4.58 LEMMA.

Let A act on R^D by the Cartesian action $f \to afa^{-1}$. Let $D = \{x_1, \ldots, x_d\}$ and $R = \{y_1, \ldots, y_r\}$. Let $S_a = \{f: f \in R^D, afa^{-1} = f\}$ and let $S_a^* = \{f: f \in R_*^D, afa^{-1} = f\}$. In the notation of the previous two paragraphs we have:

$$(1) \quad \sum_{f \in S_a} \prod_{i=1}^{d} f(x_i) = \prod_{k=1}^{d} \left(\sum_{c' \in cyc_R(a)} |c'| \cdot \left(\prod_{i \in c'} y_i \right)^{k/|c'|} X(|c'| \, | \, k) \right)^{\nu_D(a,k)}$$

$$(2) \quad \sum_{f \in S_a^*} \prod_{i=1}^{d} f(x_i) = \prod_{k=1}^{d} k^{\nu_D(a,k)} \sum_{\theta \in \mathcal{I}_k(a)} \prod_{c' \in Im(\theta)} \prod_{j \in c'} y_j \; .$$

Proof. The proof is the discussion of the previous two paragraphs extended in the obvious way to the general case. The factor $k^{\nu_D(a,k)}$ in (2) comes from the fact that if c' is associated to c and $x \in c$, then there are exactly k different choices for $f(x)$ to construct an invariant f. Once this choice is made, the value of f on all of c is determined. This choice among k values must be made for $\nu_D(a,k)$ different c' and hence the factor.

To find interesting applications of LEMMA 4.58 in the generality stated requires as much ingenuity as proving the theorem in the first place. The problem is that the expressions depend on the actual cycles c' and not just on the type of permutation a acting on R. A more tractable result is obtained if we set $y_j = 1$ for all j.

4.59 LEMMA.

If we set $y_j = 1$ for $j = 1, \ldots, r$ then LEMMA 4.58(1) and (2) become

$$(1) \quad |S_a| = \prod_{k=1}^{d} \left(\sum_{j|k} j\nu_R(a,j) \right)^{\nu_D(a,k)}$$

$$(2) \quad |S_a^*| = \prod_{k=1}^{d} k^{\nu_D(a,k)} (\nu_R(a,k))_{\nu_D(a,k)} \; .$$

Proof. Recall that $j|k$ means "j divides k." The sum in LEMMA 4.58(1) is over all c' such that $|c'|$ divides k. For each fixed j such that $j|k$, there are $\nu_R(a,j)$ such terms. This explains 4.59(1). With $y_j = 1$ for all j, LEMMA 4.58(2) involves only the $|\mathcal{I}_k|$. In general, the number of injective mappings from a set of q elements to a set of p elements is $p(p-1)(p-2) \ldots (p-q+1)$. This number is called the "falling factorial" and is denoted by $(p)_q$. By definition, $(p)_q = 0$ if $q > p$. Thus, $|\mathcal{F}_k| = (\nu_R(a,k))_{\nu_D(a,k)}$ appears in 4.59(2).

LEMMA 4.59 together with Burnside's lemma (COROLLARY 4.16(2)) gives THEOREM 4.60, a result of deBruijn.

4.60 THEOREM (deBruijn).

Let A act on R^D by the Cartesian action $f \to afa^{-1}$. Let Δ be a system of representatives for the orbits of A acting on R^D and let Δ_* be a system of representatives for A acting on the injective functions R_*^D. Then

$$(1) \quad |\Delta| = \frac{1}{|A|} \sum_{a \in A} \prod_{k=1}^{d} \left(\sum_{j|k} j\nu_R(a,j) \right)^{\nu_D(a,k)}$$

$$(2) \quad |\Delta_*| = \frac{1}{|A|} \sum_{a \in A} \prod_{k=1}^{d} k^{\nu_D(a,k)} (\nu_R(a,k))_{\nu_D(a,k)} \, .$$

In EXERCISE 4.61, we explore these results further. The references at the end of the chapter should be consulted in connection with these exercises (in particular, see the chapter by de Bruijn in the book *Applied Combinatorial Mathematics*). Also in EXERCISE 4.61 we indicate some other directions one might go in exploring the topic of orbit enumeration. There are many interesting applications of this material that make ideal topics for classroom presentations by students!

4.61 EXERCISE.

(1) Show how the identity of LEMMA 4.58(1) specializes to the Pólya action IDENTITY 4.40.
(2) Consider the case where $A = G \times H$ is the direct product of two groups G and H, G: D, H: R. Thus, for $a = (g,h)$, $f \in R^D$ is transformed into hfg^{-1}. Show that THEOREM 4.60(2) can be written

$$|\Delta_*| = P_G\left(\frac{\partial}{\partial z_1}, \ldots, \frac{\partial}{\partial z_d}\right) P_H(1 + z_1, 1 + 2z_2, \ldots, 1 + rz_r)$$

evaluated at $z_1 = \ldots = z_r = 0$.
Hint: Note that if $|\Delta_*| \neq 0$ then we must have $r \geq d$. For any integers p, q, and s, where $q \leq s$, we have

$$\left[\frac{\partial}{\partial z}\right]^q (1 + pz)^s \Big|_{z=0} = p^q s(s-1)(s-2) \ldots (s-q+1) = p^q(s)_q.$$

This latter expression occurs in THEOREM 4.60(2) with $k = p$, $q = \nu_D(a,k)$ and $s = \nu_R(a,k)$.

(3) Show that in the case where $|D| = |R|$ the identity of (2) becomes $|\Delta_*| = P_G\left(\dfrac{\partial}{\partial z_1}, \ldots, \dfrac{\partial}{\partial z_d}\right) P_H(z_1, 2z_2, \ldots, dz_d)$ evaluated at $z_1 = \ldots = z_d = 0$.

Hint: Note that $\left(\dfrac{\partial}{\partial z_1}\right)^{q_1} \ldots \left(\dfrac{\partial}{\partial z_d}\right)^{q_d} z_1^{s_1} (2z_2)^{s_2} \ldots (dz_d)^{s_d}$ is equal to

$\left(\dfrac{\partial}{\partial z_1}\right)^{q_1} \ldots \left(\dfrac{\partial}{\partial z_d}\right)^{q_d} (1+z_1)^{s_1} (1+2z_2)^{s_2} \ldots (1+dz_d)^{s_d}$ if $q_i = s_i$ for all i. Otherwise, using $|D| = |R|$, show $q_i > s_i$ for some i and both expressions are zero.

(4) Illustrate the result of EXERCISE (3) with some examples.

(5) Consider $A = G \times H$ as in EXERCISE (2) above. Show that THEOREM 4.60(1) can be written

$$|\Delta| = P_G\left(\dfrac{\partial}{\partial z_1}, \ldots, \dfrac{\partial}{\partial z_d}\right) P_H(g_1, \ldots, g_r)\Big|_{z_1 = z_2 = z_3 = \ldots = 0}$$

where $g_t = \exp\left(j \sum_t z_{t \cdot j}\right)$. The notation $z_{t \cdot j}$ means z with subscript equal to the product of t and j. Thus, $z_{2 \cdot 3} = z_6$.

(6) Consider the following two problems:

(a) The faces and the vertices of a cube are to be labeled (simultaneously) with symbols from a set R. How many ways are there to do this up to rotations of the cube?

(b) The faces of a cube are to be labeled with symbols from a set R_1 and the vertices of a cube are to be labeled with symbols from a set R_2 ($R_1 \cap R_2 = \phi$). How many ways are there to do this up to the rotations of the cube?

Discuss the solutions to these problems. Does either problem require an extension of Pólya's theorem? If so, formulate this extension in the general case.

(7) Let T and D be sets and A a group that acts on T and D. Let A act on R^D by the Pólya action. Thus, A acts on $T \times R^D$ by the Cartesian action $(a(t,f) = (at, fa^{-1}))$. Let $U_A(z_1, \ldots, z_d) = \displaystyle\sum_{t \in \Delta(T)} P_{A_t}(z_1, \ldots, z_d)$ where A_t is the stability subgroup of A at t (relative to A: T) and $\Delta(T)$ is a system of orbit representatives for A: T. Show that $U_A(r, \ldots, r)$ is the number of orbits of A acting on $T \times R^D$ where $r = |R|$. What are some combinatorial interpretations of this result? *Hint.* The collection of sets $\{\{t\} \times R^D : t \in T\}$ is a partition of $T \times R^D$. Note that the blocks of this partition are invariant under the action of the group A in the sense that $a \in A$, $a(\{t\} \times R^D) = \{at\} \times R^D$, and the latter set is still a block of the partition. This type of generalization has been developed by deBruijn (see references list at end of Part I). In fact, the idea of invariant partitions extends to any group action and was the basis

of our isomorph rejection algorithms earlier in Chapter 1 (for example, the n-queens problem of FIGURE 1.62 through FIGURE 1.77).

(8) If G acts on D and H acts on R then we have noted that the action of G × H on D × R defines an action on the sets {GRAPH(f): f ∈ R^D}. In this action, f is sent to hfg^{-1} by the element (g,h) of G × H. But, the wreath product G[H] also acts on D × R by the rule (g,φ)(i,j) = (gi,φ(i)j). Thus, G[H] also acts on the set {GRAPH(f): f ∈ R_D}. Under this action, f is sent by (g,φ) to a function whose value at x ∈ D is $\varphi(g^{-1}x)f(g^{-1}x)$. Compare these two actions on R^D by giving some examples for small order cases ($C_3[C_2]$, $C_4[C_3]$, or $S_3[S_2]$). Various applications of this action as well as a formula for this cycle index polynomial are given in the Palmer and the Robinson references cited at the end (**classical references - search Web for recent**).

(9) Consider the following figure:

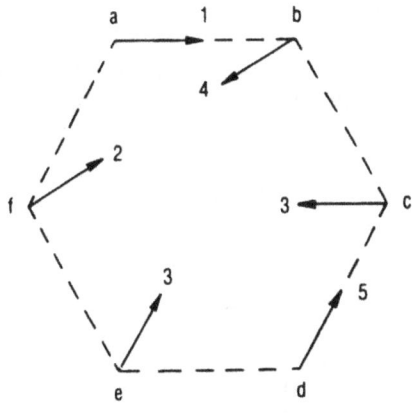

Figure 4.61

This figure shows the correspondence between functions in R^D and hexagons with unit vectors attached to each vertex. (D = {a,b,c,d,e,f}, R = 5). If one stands at a vertex of the hexagon and looks toward the center, then the other vertices are referred to as "1,2,3,4,5," left to right. Thus, at vertex a the arrow points to the first vertex left to right (which is b), at vertex b the arrow points to the fourth vertex (which is f), etc. This figure corresponds to the function f = $\begin{pmatrix} a\ b\ c\ d\ e\ f \\ 1\ 4\ 3\ 5\ 3\ 2 \end{pmatrix}$. How many such "vector labeled" hexagons are there up to rotations of the hexagon? How many up to rotations and reflections?

We conclude this TOPIC with a discussion of the constructive isomorph rejection problem. That is, we wish to actually construct a system of representatives for the orbits of any group action A: S. The reader who has not read the development of the various orbit enumeration techniques presented thus far in this TOPIC should read the material through LEMMA 4.10 and also read EXAMPLE 4.34, EXERCISE 4.35, DEFINITION 4.36, and DEFINITIONS 4.55 and 4.56. We have already presented the most basic techniques for solving isomorph rejection problems in EXERCISE 1.38 and in connection with the discussion of the solution to the n-queens problem (FIGURES 1.62–1.77). We now discuss a more specialized but very useful technique called the method of "orderly algorithms." A graph theoretic applications of this method are presented independently in Part II (items 6.67, 6.76–6.79 of study guide). They are in BASIC CONCEPTS Chapter 6. Our approach to orderly algorithms will emphasize a particular but very important case, the generation of set partitions of a fixed type. We first consider a class of problems where such set partitions arise naturally. If you wish, you may look quickly at the material from here through EXERCISE 4.64 and then return to this material for careful study in connection with EXERCISE 4.70(3).

We consider the group A of rotations and reflections of the square as shown in FIGURE 4.2. Let R^D be the set of functions with domain $D = \{a,b,c,d\}$ and range $R = \{L,M,R\}$. A function $f \in R^D$ will be called an "LMR-diagram" as shown in FIGURE 4.62.

4.62 LMR-DIAGRAM FOR VERTICES OF A SQUARE.

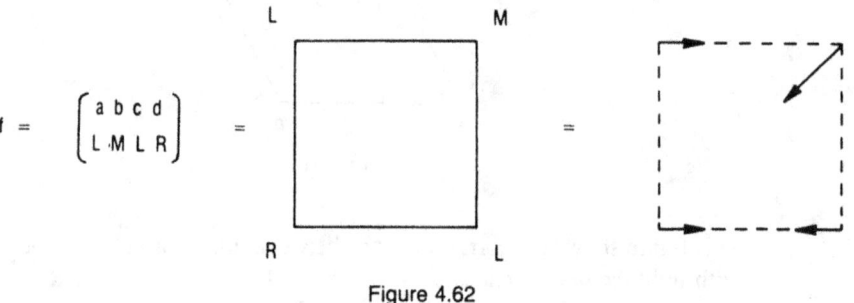

$$f = \begin{pmatrix} a & b & c & d \\ L & M & L & R \end{pmatrix}$$

Figure 4.62

The function f of FIGURE 4.62 corresponds to the labeling of the vertices of a square with unit vectors. If $f(a) = L$, then as one stands at vertex a and faces the center of the square, a unit vector is drawn pointing to the left. If $f(a) = M$ the vector is drawn towards the middle, and if $f(a) = R$, the vector is drawn towards the right as shown in FIGURE 4.62. This idea was explored in a slightly more complex situation in EXERCISE 4.61(9). Associated with each function is its coimage (NOTATION 1.6). The coimage(f) is the set partition of D defined by $\{f^{-1}(x): x \in image(f)\}$. For the function f of FIGURE 4.62, we have image(f) = $\{L,M,R\}$ and hence coimage(f) = $\{f^{-1}(L),f^{-1}(M),f^{-1}(R)\}$ = $\{\{a,c\},\{b\},\{d\}\}$. A rotation τ acts on an L,M,R diagram by rotating the structure 90° counterclockwise ($f \rightarrow \tau f \tau^{-1}$) and does not change the L, M, or R labels

42

(only their positions). A reflection ρ acts by the rule $f \rightarrow \rho f \rho^{-1} = \rho f \rho$ and changes L to R and R to L. Thus, the reflection $\rho_r = (b,d)$ applied to the function f of FIGURE 4.62 produces the function $g = \begin{pmatrix} a & b & c & d \\ R & L & R & M \end{pmatrix}$. The action of A on R^D is the Cartesian action on functions of DEFINITION 4.56.

If a group A acts on a set D, then A acts on the set $\Pi(D)$ of all partitions of D in an obvious way. If $\mathscr{C} = \{B_1, B_2, \ldots, B_p\}$ is a partition of D with blocks B_s, and if $a \in A$ then $a\mathscr{C} = \{aB_1, aB_2, \ldots, aB_p\}$ where aB_s is the set obtained by applying a to each element of B_s. Let $\nu(\mathscr{C}, j)$ denote the number of blocks B_t of $\mathscr{C}$ with $|B_t| = j$. The vector $(\nu(\mathscr{C}, 1), \nu(\mathscr{C}, 2), \ldots, \nu(\mathscr{C}, d))$ is called the *type* of the partition $\mathscr{C}$. This is sometimes an awkward notation (suppose $d = 20$, $\nu(\mathscr{C}, 10)$ = 2, then the type is $(0,0,0,0,0,0,0,0,0,2,0,0,0,0,0,0,0,0,0,0))$. Another notation is $1^{\nu(\mathscr{C}, 1)} 2^{\nu(\mathscr{C}, 2)} \ldots d^{\nu(\mathscr{C}, d)}$ where an expression of the form k^0 is always omitted. Thus, 10^2 would denote a partition of 20 with two blocks of size 10, $1^3 2^4 5^2$ would denote a partition of 21 with three blocks of size 1, four blocks of size 2, and two blocks of size 5.

If we wish to construct a system of orbit representatives for the Cartesian action of a group A acting on a set of functions R^D, then we may first construct a system $\Delta(\Pi)$ of representatives for A acting on the partitions of D (i.e., $\Pi(D)$). If $\mathscr{C} \in \Delta(\Pi)$ then let $A_{\mathscr{C}}$ denote the stability subgroup of A at $\mathscr{C}$ (all $a \in A$ such that $a\mathscr{C} = \mathscr{C}$). We may then construct a system of orbit representatives for $A_{\mathscr{C}}$ acting on the set of all functions $f \in R^D$ such that coimage$(f) = \mathscr{C}$. The resulting set will be a system of representatives for $A: R^D$. This process is illustrated in FIGURE 4.63. The general idea is developed in EXERCISE 4.64(3) and relates to the example of this paragraph by taking $S = R^D$ and $\mathscr{F} = \Pi(D)$. For $i \in \Pi(D)$, F_i of EXERCISE 4.64(3) is all functions in $R^D = S$ with coimage equal to i. For $a \in A$, $F_i \in \mathscr{F}$, $a F_i = F_{ai}$ where ai is defined as in the previous paragraph (take $i = \mathscr{C}$ there).

In FIGURE 4.63, a system of representatives $\Delta(\Pi)$ for the dihedral group D_8 acting on the partitions of the set $\{a,b,c,d\}$ is shown. The symbols a, b, c, and d refer to the vertices of the square of FIGURE 4.2. The set $\Delta(\Pi)$ is obtained by a simple inspection in this case but will be obtained below by a specific algorithm for the general case. For each $\mathscr{C}$ in $\Delta(\Pi)$ we compute the stability subgroup $A_{\mathscr{C}}$. For $\{\{a\}\{b\}\{c\}\{d\}\} = \mathscr{C}$ there are no LMR diagrams with $\mathscr{C}$ as coimage. For $\{\{a\ b\}\{c\}\{d\}\} = \mathscr{C}$, $A_{\mathscr{C}} = \{e, \rho_s\}$. For $\{\{a\ c\}\{b\}\{d\}\} = \mathscr{C}$, $A_{\mathscr{C}} = \{e, \rho_r, \rho_t, \tau^2\}$. For $\{\{a\ c\}\{b\ d\}\}$, the stability subgroup is all of D_8. For $\{\{a\ b\ c\}\{d\}\}$, the stability subgroup is $\{e, \rho_t\}$ and for $\{a\ b\ c\ d\}$ the stability subgroup is again D_8. For each $\mathscr{C}$ we list a system of representatives for the set of LMR diagrams with coimage equal to $\mathscr{C}$ under the action of $A_{\mathscr{C}}$. This set is obviously complete for $A: R^D$. In general, this method defines a recursive algorithm for isomorph rejection in the case of the Cartesian action of A on R^D. For each $\mathscr{C}$, the problem is reduced to an isomorph rejection problem for the Cartesian action of A on

$R_*^{\mathscr{C}}$, the injective functions from $\mathscr{C}$ to R. In general, each of these subproblems may be treated recursively in the same manner.

4.63 ORBIT REPRESENTATIVE SYSTEMS FOR LMR DIAGRAMS.

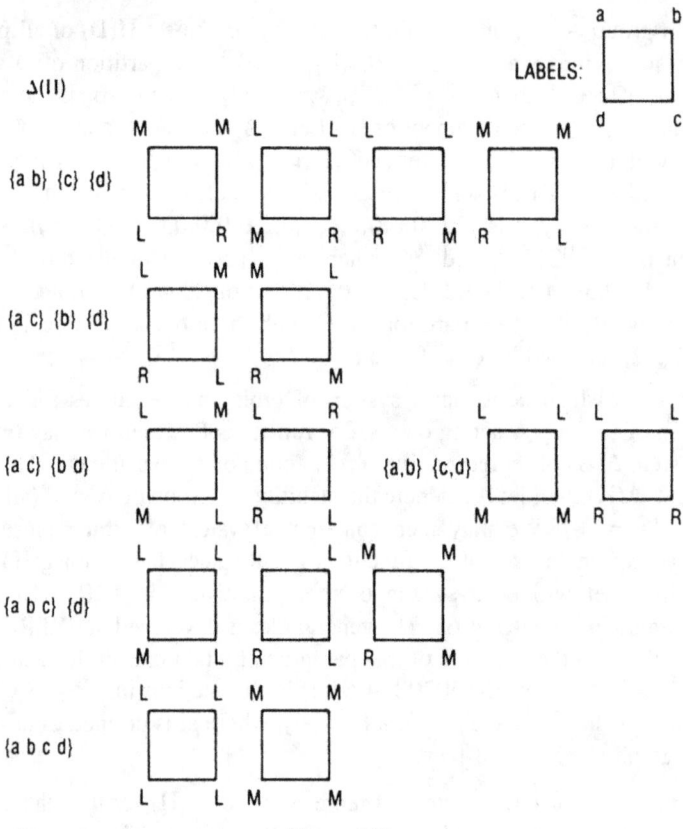

Figure 4.63

4.64 EXERCISE.

(1) Let A be a group that acts on D and hence on R^D by the Pólya action (DEFINITION 4.36). Let D = {a,b,c,d}, R = {$y_1, y_2, \ldots, y_r$}, A = D_8 as in FIGURE 4.63. Using $\Delta(\Pi)$ of FIGURE 4.63, describe for each $\mathscr{C} \in \Delta(\Pi)$ the system of orbit representatives for $A_{\mathscr{C}}$ acting on the functions $f \in R^D$ with coimage(f) = $\mathscr{C}$. For each $\mathscr{C}$, express the number of such orbit representatives as a function of r and, by summing over all $\mathscr{C}$ in $\Delta(\Pi)$, express the total number of orbit representatives for A acting on R^D as a function of r. Compare this result with the formula obtained by applying PÓLYA'S THEOREM 4.43 to the same problem.

(2) Work EXERCISE 4.64(1) above where D = {a,b,c,d,e,f} are the vertices of a hexagon and A is the dihedral group (all rotations and reflections). Suppose that "side conditions" are placed on the structures being listed such as (a) if any symbol y_i of R appears at a vertex of the hexagon it must also appear at at least one other vertex, and (b) no two adjacent vertices of the hexagon are labeled with the same symbol of R. Describe the orbit representatives that satisfy conditions (a) and (b). Can PÓLYA'S THEOREM 4.43 or BURNSIDE'S LEMMA 4.15, COROLLARY 4.16 be applied to this situation? Explain.

(3) Let S be a set, and let A be a group that acts on S. Let $\mathcal{F} = \{F_i : i \in \mathcal{I}\}$ be a partition of S (the set $\mathcal{I}$ is an "index set"). If for all i ∈ $\mathcal{I}$ and all a ∈ A, $aF_i = F_j$ for some j ∈ $\mathcal{I}$, then $\mathcal{F}$ is called an A-*invariant partition* or simply an *invariant partition*. In this case, we say that A acts on the partition $\mathcal{F}$. Let $\Delta(\mathcal{F})$ be a system of orbit representatives for A: $\mathcal{F}$ and let $\Delta(F)$ be a system of orbit representatives for A_F acting on F where F is a block of $\mathcal{F}$ and A_F is the stability subgroup of A at F (all a ∈ A, aF = F). Let $\Delta(S)$ be the union over all F in $\Delta(\mathcal{F})$ of $\Delta(F)$. Prove that $\Delta(S)$ is a system of orbit representatives for A acting on S. What are $\mathcal{I}$, $\Delta(\mathcal{F})$, and $\Delta(F)$ for the example of FIGURE 4.63?

(4) Let G and H be groups and G × H their direct product. Let G × H act on R^D by the Cartesian action $((g,h)f = hfg^{-1})$. If H = PER(R) is the symmetric group on R, find a natural bijection between the orbits of G × H acting on R^D and the orbits of G acting on $\Pi(D)$, the set of all partitions of D.

4.65 STRUCTURE OF ORDERLY ALGORITHM 4.66:
The orderly map B.

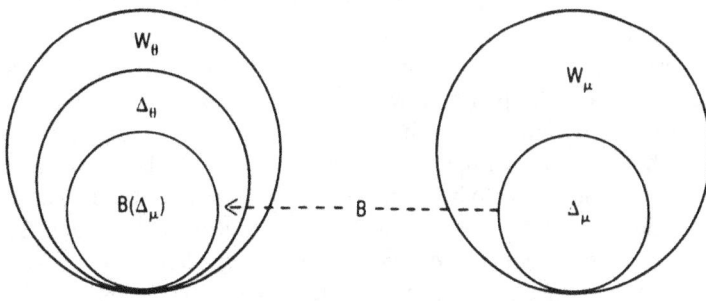

Figure 4.65

45

As EXERCISE 4.64 indicates, certain basic combinatorial isomorph rejection problems are central to a number of related problems. The generation of set partitions by type under a group action is such a problem. Independent of any group theoretic considerations, the problem we are dealing with is, given W and a subset Δ, find Δ. We assume that, given an element of W, we can test it to see if it is an element of Δ. Suppose that $\{W_\theta : \theta \in \Theta\}$ is a partition of W where Θ is a linearly ordered set. For each block W_μ, let $\Delta_\mu = W_\mu \cap \Delta$. An *orderly map* B for this partition is a function defined on blocks W_μ, $\mu \neq$ first element of Θ, which maps W_μ to W_θ where θ is the predecessor of μ in the linear order on Θ. In addition, B must satisfy the condition $B(\Delta_\mu) \subseteq \Delta_\theta$. The situation is illustrated in FIGURE 4.65. A number of useful algorithms have the general structure of ALGORITHM 4.66.

4.66 ORDERLY ALGORITHM WITH ORDERLY MAP B.

procedure   FIND Δ.
    $\theta := $ first element of Θ;
    $U := \Delta_\theta$
    while $\theta \neq$ last element of Θ *do*
        begin
            $\mu := $ successor of θ in Θ;
            compute $B^{-1}(\Delta_\theta)$;
            find $B^{-1}(\Delta_\theta) \cap \Delta$;
            $\Delta_\mu := B^{-1}(\Delta_\theta) \cap \Delta$;
            $U := U \cup \Delta_\mu$;
            $\theta := \mu$;
        end
    $\Delta := U$;

The intuitive idea of ALGORITHM 4.66 is that one starts with the first element θ of Θ and constructs, in a manner unspecified, Δ_θ. One then computes $B^{-1}(\Delta_\theta)$. The condition $B(\Delta_\mu) \subseteq \Delta_\theta$, μ the successor of θ, assures us that all elements of Δ_μ are contained in $B^{-1}(\Delta_\theta)$. One then searches the elements of $B^{-1}(\Delta_\theta)$ for all elements of Δ to obtain Δ_μ. This process is repeated until θ becomes the last element of Θ. At this point all elements of Δ have been found. This process as a general method is very simple. The challenge comes in constructing, for particular cases, the partition W_θ and the map B to minimize the amount of testing to see if elements are in Δ.

We shall present a number of examples. One wants the construction of Δ_θ for the first element θ of Θ to be trivial. The construction of the elements of $B^{-1}(\Delta_\theta)$ should be easy for all θ and the difference between $B^{-1}(\Delta_\theta)$ and Δ_μ should be small. These tradeoffs present some interesting challenges. It is obvious that the idea of the orderly algorithm can be extended to cases where Θ is not a linearly ordered set but any partially ordered set. For example, Θ might be the vertex set of a rooted tree. The algorithm would start with Δ_θ given, θ the root of the tree. The map B in this case would map W_μ for all μ, $\mu \neq$ root, to W_θ, where θ is the farther of μ. For the orderly algorithm one first computes $B^{-1}(\Delta_\theta)$ where θ is the root and searches for the sets Δ_μ for each son μ of θ. The process is then repeated on each subtree of the root. We shall illustrate this approach with an example rather than formally describe the method in this generality (see FIGURE 4.69 and related discussion).

4.67   ORDERLY ALGORITHM FOR RANGE ACTIONS.

Let $W = \bigcup\limits_{d=1}^{p} \underline{r}^{\underline{d}}$. Let H be a group acting on $\underline{r}$. Then H acts on W by the rule $f \to hf$, for $f \in W$, $h \in H$. Let $W_d = \underline{r}^{\underline{d}}$ and $\Theta = \underline{p}$. Using one line notation, let $f = (f_1, \ldots, f_d)$ and define $B(f) = (f_1, \ldots, f_{d-1})$. Let Δ_d be a lexicographically minimal system of orbit representatives for H acting on $\underline{r}^{\underline{d}}$. Then it is easily seen that $B(\Delta_d) \subseteq \Delta_{d-1}$ (EXERCISE 4.68(1)). Hence, B is an orderly map in the sense of FIGURE 4.65. In order to be more specific, take $r = 4$ and $H = D_8$, the dihedral group. H may be thought of as the permutations of the vertices of the square

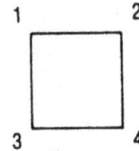

Figure 4.67

resulting from all rotations and reflections of the square (as in the case of FIGURE 4.2 with {a,b,c,d} replaced by {1,2,3,4}). In this case, $\Delta_1 = \{1\}$. $B^{-1}(\Delta_1) = \{(1,1),(1,2),(1,3),(1,4)\}$. We must check to see if each of these elements of $\underline{4}^{\underline{2}}$ is minimal in its orbit under the action of H. Consider (1,4) for example. For $h \in H$, $h(1,4) = (h1,h4)$. Clearly, if (h1,h4) is less than (1,4) in lex order, then $h1 = 1$. This means that h is the identity or the reflection about the line joining vertex 1 to vertex 3 in the above figure. For this reflection, $h4 = 2$, so (1,4) is not lex minimal in its orbit and is thus not an element of Δ_2. The elements (1,1), (1,2), and (1,3) are all in Δ_2 as they are easily seen to be lex minimal. For simplicity, we write $\Delta_2 = \{11,12,13\}$. Then $B^{-1}(\Delta_2) = \{111,112,113,\underline{114},121,122,123,124,131,132,133,\underline{134}\}$ where the elements not in Δ_3 are indicated.

4.68   EXERCISE.

(1) Prove that the map B of EXAMPLE 4.67 is an orderly map (satisfies $B(\Delta_d) \subseteq \Delta_{d-1}$).

(2) Extend EXAMPLE 4.67 to the general case where $H = D_{2r}$ is the dihedral group acting on the set $\underline{r}$. This may be thought of as the permutations of the vertices of a regular polygon with r vertices resulting from all rotations and reflections. Describe Δ_d for all d.

(3) Let $W = \underline{r}^{\underline{d}}$ and let $W_k = \left\{ f: \sum_{i=1}^{d} f(i) = k \right\}$. Let $\Theta = (dr, dr-1, \ldots, d)$. Write $f = (f_1, f_2, \ldots, f_t, r, \ldots, r)$ where $f_t < r$, $t \leq d$. Define the orderly map B by the rule $B(f) = (f_1, \ldots, f_t+1, r, \ldots, r)$. Let Δ be a lexicographically minimal system of orbit representatives for the Pólya action of G on $\underline{r}^{\underline{d}}$ (DEFINITION 4.36). Prove that $B(\Delta_k) \subseteq \Delta_{k+1}$ and hence that B is an orderly map. (Note that k is "larger" than $k+1$ in the order on Θ.) Give some examples of the resulting algorithm.

We now give an example of an orderly algorithm where the index set for the blocks of the partition of W has the structure of an ordered rooted tree. We consider the important case of the generation of set partitions by type discussed above. Consider FIGURE 4.69. FIGURE 4.69(a) is a binary tree whose vertices are integral partitions of 6. We use the mixed notation $(\tau_1, \ldots, \tau_j, 1^k)$ to indicate the integral partition $(\tau_1, \tau_2, \ldots, \tau_j, 1, \ldots, 1)$ of d where the entries are in nonincreasing order and there are k ones. The general rule for constructing the tree is given in FIGURE 4.69(b). If the rule for constructing a son of a vertex is not applicable, then that vertex is labeled by ø. The terminal nodes of the tree are either the symbol ø or an integral partition without 1's. The reader is asked, in EXERCISE 4.70(1), to show that this method, in fact, generates all partitions of d. We call the tree of FIGURE 4.69 the *type tree of order* d (order 6 is shown in FIGURE 4.69(a)).

4.69 THE TYPE TREE.

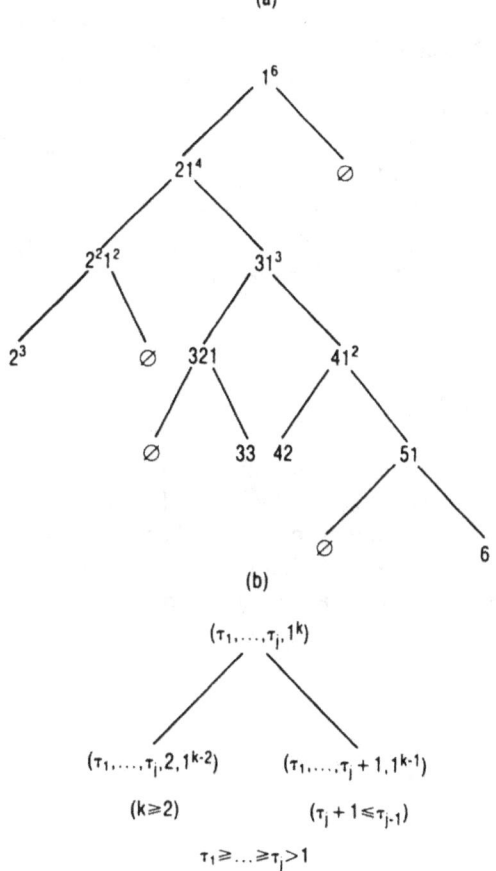

Figure 4.69

Let Θ denote the vertices of the type tree. Let W be all set partitions of $\underline{d}$ and let W_θ, $\theta \in \Theta$ be all partitions of type θ. If G is a group that acts on $\underline{d}$, then G acts on W as explained in connection with FIGURES 4.62 and 4.63. We say that a partition $\mathscr{C} = \{B_1, \ldots, B_p\}$ of $\underline{d}$ is *in order* if $i < j$ implies that $|B_i| > |B_j|$ or, if $|B_i| = |B_j|$, then the minimum element of B_i is less than the minimum element of B_j. If $\mathscr{C}$ has its blocks in order, and each block is written in increasing order, then the sequence of integers obtained by concatenating these blocks will be called the *standard form* of $\mathscr{C}$. For example, the partition $\mathscr{C} = \{\{3,5,8\},\{1,4\},\{2,6\},\{7\}\}$ is in order and the blocks are in increasing order. The standard form of this partition is 3,5,8,1,4,2,6,7.

It is obvious that given the type and the standard form of a partition, one can construct the partition. The reader should practice applying some permutations to partitions in standard form. For instance, suppose that $\sigma = (1,3,5,7)(2,4,6,8)$

is a permutation of $\underline{8}$ in cycle notation. Let $\mathscr{C}$ be as above, then $\sigma\mathscr{C}$ is 2,5,7,3,6,4,8,1 in standard form. We order the sets W_θ lexicographically in standard form. For each $\theta \in \Theta$ let Δ_θ be a minimal system of orbit representatives with respect to this lex order. We now define, for each θ, not the root of the type tree, the map $B: W_\theta \to W_\mu$ where μ is the father of θ in the type tree. Let $\mathscr{C} = \{B_1, . . .,B_p\}$ be in order. Let j be the largest index such that $|B_j| > 1$. Define $B(\mathscr{C})$ to be the partition obtained from $\mathscr{C}$ by removing the largest element from B_j and creating from that element a new block of size 1. We ask the reader to show in EXERCISE 4.70(2) that $B(\Delta_\theta) \subseteq \Delta_\mu$, and hence that B is an orderly map. We call B the *type tree map*. The reader is asked in EXERCISE 4.70(3) to describe and give examples of the resulting orderly algorithm.

4.70 EXERCISE.

(1) Prove that every integral partition of d appears just once as a vertex of the type tree of order d, defined by FIGURE 4.69.
(2) Prove that the type tree map B defined above is an orderly map.
(3) Describe the orderly algorithm based on the type tree map B of (2). Give some examples of applications to the problems of EXERCISE 4.64.

Classical References

Combinatorial Mathematics:

deBruijn, N.G., "Pólya's Theory of Counting," *Applied Combinatorial Mathematics*, Beckenbach, E.F. (Editor), Wiley, New York, 1964.

Other related articles follow:

deBruijn, N.G., *Enumerative combinatorial problems concerning structures*, Nieuw. Arch. Wisk. *11* (1963), 142–161.

deBruijn, N.G., and Klarner, D.A., *Enumeration of generalized graphs*, Indag. Math. *31*(1) (1969), 1–9.

deBruijn, N.G., *A survey of generalizations of Pólya's enumeration theorem*, Nieuw. Arch. Wisk. *19* (1971), 89–112.

The proof of White's lemma is based on the following paper:

White, D.E., *Counting patterns with a given automorphism group*, Proc. AMS *47* (1975), 41–44.

For some related references take a look at the following articles:

Sheehan, J., *The number of graphs with a given automorphism group*, Can. J. Math. *20* (1968), 1068–1076.

Stockmeyer, P.K., "Enumeration of graphs with prescribed automorphism group," Ph.D. Thesis, Univ. of Michigan, Ann Arbor, Mich., 1971.

White, D.E., *Classifying patterns by automorphism group: An operator theoretic approach*, Discrete Math. *13* (1975), 277–295.

For an introduction to the applications of Burnside's lemma to chemistry, see:

McLarnan, T.J., and Moore, P.B., "Graph-Theoretic Enumeration of Structure Types: A Review," in *Structure and Bonding in Crystals* II, O'Keeffe, M., Navrotsky, A. (Editors), Academic Press, New York, 1981.

McLarnan, T.J., *Mathematical tools for counting polytypes*, Zeitshrift für Kristallographie *155* (1981), 227–245. (See also, pp. 247–268 and pp. 269–291.)

The first systematic treatment of the subject of enumerating orbits of group actions was by Redfield and was reported in:

Redfield, J.H., *The theory of group-reduced distributions*, American J. Math. *49* (1927), 433–455.

Little notice was taken of Redfield's paper until after the following famous paper by Pólya:

Pólya, G., *Kombinatorische Ansahlbesstimmungen für Gruppen, Graphen, und Chemishe Verbindungen*, Acta Math. *68* (1937), 145–254.

Several of the combinatorial mathematics books in the list of texts in this reference list contain treatments of Pólya's theorem. Numerous examples are given in the books by Tucker, Liu, and Cohen.

One area where Pólya's theorem has been applied extensively is in graph theory. For more information refer to the following text:

Harary, F., and Palmer, E.M., *Graphical Enumeration*, Academic Press, New York, 1973.

For an interesting application of wreath products, we refer the reader to the following article:

Hanlon, P., *A cycle index sum inversion theorem*, J. Comb. Theory *30* (1981), 248–269.

The action of the wreath product described in EXERCISE 4.61(8) is discussed, together with some applications, in:

Palmer, E.M., and Robinson, R.W., *Enumeration under two representations of the wreath product*, Acta Math. *131* (1973), 123–143.

Orbit enumeration problems play a basic role in multilinear algebra. The reader interested in extensions of Burnside's lemma and Pólya's theorem along these lines might look at the following:

Merris, R., *Generalized matrix functions and pattern inventory*, Linear and Multilinear Algebra v. 12, 1983, 315–327.

White, D.E., *Multilinear techniques in Pólya enumeration theory*, Linear and Multilinear Algebra 7 (1979), 299–315.

Williamson, S.G., *Isomorph rejection and a theorem of deBruijn*, SIAM J. Comp. 2(1973), 44–59.

The basic idea of an "orderly algorithm" seems to have occurred to a number of people working on applied problems in chemistry, engineering, and computer science. The first careful statement of the general method seems to be due to Read in the interesting paper

Read, R.C., *Every One a Winner or How to Avoid Isomorphism Search When Cataloguing Combinatorial Configurations*, Annals of Discrete Math. *2* (1978), 107–120.

52

Index

NOTES

NOTES

NOTES

NOTES